T0281200

Our Universe can be described mathematically by a simple model developed in 1922 at Petrograd (St. Petersburg) by Alexander Friedmann (1888–1925), who predicted, before there was any observational evidence, that the whole Universe would expand and evolve with time. He was an outstanding Soviet physicist, and this vivid biography is set against a wide historical background. The book is a window on his school and university years, military service, and teaching and research during a seminal period of Soviet history. The authors include unique archival material, such as Friedmann's letters from the Front, as well as contemporary records and reminiscences of colleagues. There is a detailed treatment of his work in theoretical cosmology (1922–24), set in the context of the organization of Soviet science at the time.

ALEXANDER A. FRIEDMANN:
THE MAN WHO MADE THE UNIVERSE EXPAND

ALEXANDER A. FRIEDMANN: THE MAN WHO MADE THE UNIVERSE EXPAND

EDUARD A. TROPP
VIKTOR Ya. FRENKEL
and
ARTUR D. CHERNIN

Translated by
ALEXANDER DRON and MICHAEL BUROV

CAMBRIDGE
UNIVERSITY PRESS

CAMBRIDGE UNIVERSITY PRESS
Cambridge, New York, Melbourne, Madrid, Cape Town, Singapore, São Paulo

Cambridge University Press
The Edinburgh Building, Cambridge CB2 2RU, UK

Published in the United States of America by Cambridge University Press, New York

www.cambridge.org
Information on this title: www.cambridge.org/9780521384704

© Cambridge University Press 1993

First published 1993
This digitally printed first paperback version 2006

A catalogue record for this publication is available from the British Library

Library of Congress Cataloguing in Publication data
Tropp, E. A. (Eduard Abramovich)
[Aleksandr Aleksandrovich Fridman. English]
Alexander A. Friedmann: the man who made the universe expand / by
Eduard A. Tropp, Viktor Ya. Frenkel, and Artur D. Chernin;
translated by Alexander Dron and Michael Burov,
p. cm.
Includes bibliographical references and index.
ISBN 0-521-38470-2
1. Fridman, A. A. (Aleksandr Aleksandrovich), 1888–1925.
2. Cosmology. 3. Physicists – Soviet Union – Biography.
4. Astrophysics – Soviet Union – Biography. I. Frenkel, Viktor IAkovlevich.
II. Chernin, A. D. (Artur Davidovich). III. Title.
QC16.F73T7613 1993
530'.092–dc20 [B] 92-28315 CIP

ISBN-13 978-0-521-38470-4 hardback
ISBN-10 0-521-38470-2 hardback

ISBN-13 978-0-521-02588-1 paperback
ISBN-10 0-521-02588-5 paperback

Contents

Preface   *page* ix

1 The Friedmanns and the Voyacheks   1

2 At the 2nd St. Petersburg Gymnasium   12

3 University years, 1906–14   28

4 In search of a way   58

5 War years   68

6 Moscow – Perm – Petrograd   86

7 Theoretical department of the Main Geophysical Observatory   97

8 Space and time   114

9 Geometry and dynamics of the Universe   144

10 Petrograd, 1920–24   176

11 The final year   194

12 Friedmann's world   215

Conclusion   254

Main dates in Friedmann's life and work   256

Bibliography   258

Name index   262

Preface

This book came out in Russian in 1988, the centenary of the birth of Alexander Friedmann, an outstanding Soviet scientist. The anniversary was widely marked by the scientific community with national and international conferences and symposia dedicated to this event as a tribute to the enormous contribution A. A. Friedmann made to the development of hydrodynamics, meteorology and, particularly, relativistic cosmology. Very little has been written about Friedmann. The present book is the first detailed biography of the scientist. The material is generally given chronologically and is based on the scientific works of Friedmann, documentary records and reminiscences by his contemporaries published in the late 1920s. The book also includes unpublished reminiscences about Alexander Friedmann.

The division of the work between the authors was as follows: Chapters 1–3, 5, 6 and 10, as well as the last section of Chapter 9, have been written by V. Ya. Frenkel; Chapters 4, 7 and 11 by E. A. Tropp; Chapters 8, 9 and 12 by A. D. Chernin. Given this clear division, in some chapters the material is presented in the first person singular. As to the method of citation, given that the book is popular science, the authors decided not to overburden it with footnotes. Quotations are either from archives or from the literature given at the end of the book.

The authors would like to thank the staff workers of the Leningrad archives for their assistance. We are grateful to Professor G. A. Grinberg and Professor L. G. Loitsyansky, and also to S. Ye. Malinina, O. N. Trapeznikova and A. B. Shekhter who shared their reminiscences with us. We owe much important information to Ye. B. Belodubrovsky, A. S. Korovchenko, L. V. Rukhovets, Ye. S. Selezneva and I. I. Shakura. We would like to thank Professor A. A. Grib and Professor V. V. Ivanov for reading the manuscript and making valuable suggestions.

We are glad that the book is finally coming out in English. We would like to express our sincere gratitude to Professor M. Demianski and Professor M. S. Longair whose kind attention made this English edition possible.

1

The Friedmanns and the Voyacheks

Do we know much about our ancestors? How little we know about them! Moscow schoolchildren have been reported to remember at best the names of their grandfathers. Adults who care to think about their roots know the names of their great-grandparents and the names and patronymics of their grandparents.

It is different with people who have left their mark in politics, science or culture. So, what is known about the hero of this book – Alexander Alexandrovich Friedmann? A sample opinion poll among physicists showed that his works of 1922–24 on relativistic cosmology are included among the two or three most outstanding achievements of Soviet physicists. It would seem that interest in his personality would be heightened and therefore satisfied. Yet, it is not the case. His biography was published only once, in a thin brochure put out by Znanie (Knowledge) Publishers. The massive volume of *Classics of Science*, devoted to Friedmann and published in 1966, contains his short autobiography – "Curriculum Vitae." It also contains Friedmann's major works in hydromechanics, dynamic meteorology, atmospheric physics and relativistic cosmology. The "Addenda" to the volume have a few reminiscences of his contemporaries about him, reprinted mainly from journals and magazines of the 1920s. There are also Friedmann's extremely interesting letters to Vladimir Steklov, his teacher, whom he so much revered, and to Boris Golitsyn – two outstanding Russian scientists with whom he was associated for many years. These materials are a source for biographical notes on Friedmann. No biographical dictionary of science or Soviet encyclopaedia fails to mention his name, starting with the first edition of the Great Soviet Encyclopaedia, in the preparation of which he himself took part.

A few of Friedmann's letters to P. S. Ehrenfest, an outstanding physi-

1

cist who worked in four countries – Austria, Germany, Russia and the Netherlands – have been published. They reflect Friedmann's contacts with Ehrenfest in 1908–12, when a modern physics circle headed by Ehrenfest was working in St. Petersburg. That is all there is, or in a more cautious vein – almost all ... Not much, of course, considering what Friedmann accomplished.

The available fragmentary materials tell us that Friedmann's father, also Alexander Alexandrovich Friedmann, was an artiste and musician. His mother, Lyudmila Ignatievna Voyachek (Friedmann) was a conservatory-educated pianist and music teacher. The name Voyachek is fairly well known, particularly in Leningrad, Friedmann's home town. Vladimir Ignatievich Voyachek (1876–1971) was a famous otolaryngologist, a full member of the Academy of Medical Sciences, a Hero of Socialist Labor (the highest award in the USSR in peace time), and Lieutenant-General of the Military Medical Service. He headed the clinic of the S. M. Kirov Military Medical Academy. The clinic was named after Voyachek in his lifetime and exists today. On the eve of the 50th anniversary of Friedmann's article on relativity, one of the authors of this book (V. F.), preparing an article dedicated to this event, visited Voyachek. Vladimir Ignatievich was 94 at the time, and he shared some information about his ancestors. His father, Friedmann's grandfather, Ignaty Kasparovich Voyachek, was a musician and composer who came to Russia from Moravia.[1] There is a short entry about him in the Soviet Musical Encyclopaedia – he worked in the Mikhailovsky Theater and later in the Marlinsky Theater in St. Petersburg. The archival records on theaters are kept in the "Imperial Collection," which gave grounds for hope that his detailed file might be found. There is also information about I. K. Voyachek in H. Riemann's famous pre-revolutionary *Dictionary of Music*, as well as in the *Encyclopedic Dictionary* of Brockhaus and Efron. Most unexpectedly, Riemann's dictionary had an entry on Friedmann's father! As it turned out, in addition to being a performer at Imperial theaters, he composed music for two ballets and an overture, having completed a conservatory course in composition taught by Rimsky-Korsakov. His file is also to be found in the Imperial Collection. One could also hope to find in this collection some information about Friedmann's grandfather on his father's side – he worked as a doctor's assistant in the St. Petersburg Palace Administration. So one could expect that the collections of Leningrad archives – the Central State Historical Archives of the USSR in Senate Square, the State Historical Archives of Leningrad in quiet

[1] His original name was Ignác František Vojáček.

Pskovsky Street, the Leningrad section of the Archives of the USSR Academy of Sciences on Universitetsky Embankment – would contain many interesting materials about Friedmann the scientist. In what field of science did he specialize? According to the biographical dictionaries published in Kiev by Naukova Dumka Publishers, he specialized in mathematics and/or mechanics ("Mathematicians and Students of Mechanics"), in astronomy ("Astronomers"), and in physics ("Physicists"). In one capacity out of three (or even four) – this is the way he is known to historians of science.

The entries in the dictionaries begin in the same way: "Alexander Alexandrovich Friedmann was born in St. Petersburg on June 29, 1888." There is the usual question, though, whether it was the New or Old Style. Only for those born after the October Revolution is there no such question – the New Style for them!

Two additional weeks – is it really so important? But historians are fond of establishing exact dates, and a correction of the established version always gives some satisfaction to researchers, and the longer the time distance from the newly established to the old wrong date, the greater the satisfaction.

Friedmann's above-mentioned "Curriculum Vitae" written by him begins like this:

Alexander Alexandrovich Friedmann was born in St. Petersburg on June 17, 1888; in 1897 he was admitted to the preparatory class of the 2nd Gymnasium [a kind of high school designed to prepare students for university education], which he left in 1906 with a gold medal.

Selected Works, p. 386

It may be noted in passing that $17 + 12 = 29$, which justifies June 29 given in the above-mentioned dictionaries . . . The records of the St. Petersburg gymnasia are well preserved in the State Historical Archives of Leningrad (SHAL) in Pskovsky Street. As a matter of fact, the files on the 2nd Gymnasium are kept in the extensive collection No. 174 containing the file of Friedmann, Jr. including, specifically, the following certificate:

CERTIFICATE
In the consistorial birth registers of the Vvedensky Church [the Church of the Presentation of the Blessed Virgin in the Temple] of the Semenovsky Life-Guard Regiment for 1888, in the first section concerning births, under item No. 182 of the male sex, appears:

To Alexander Alexandrovich Friedmann, performer in the ballet company of the Imperial St. Petersburg theaters, and his wife Lyudmila Ignatievna, both of Orthodox faith, in the year one thousand eight hundred and eighty-eight was born on the fourth day of June and baptized on the twenty-ninth day of June a son Alexander.

The godparents were: Alexander Alexandrovich Oblakov, performer in the ballet company of the Imperial St. Petersburg theaters, and Maria Alexandrovna Friedmann, daughter of a provincial councillor.

In witness whereof the official seal of the aforesaid church of the Semenovsky Regiment is affixed by

Archpriest Sergei Bogoyavlensky
Psalm-reader Ivan Fedorov

January 26, 1889, St. Petersburg

So, Friedmann was born on June 4; 4 + 12 = 16 – there is an evident inaccuracy in Friedmann's "Curriculum Vitae" (even though, in one of his persons, he was a mathematician; but, may we add, involved in approximate calculations, and approximation to the truth by one day is quite sufficient).

So let us trust Sergei Bogoyavlensky who emerges here from obscurity and whose signature is confirmed by the senior psalm-reader Ivan Fedorov (let us note that Friedmann's father was a dancer in a ballet company and his godparents were, on one side, his father's colleague from the same company and, on the other side, evidently, his father's sister). Let us now give some biographical notes on "our" Alexander Alexandrovich Friedmann's grandfathers (and a little about the grandmothers) and about his parents.

Alexander Ivanovich Friedmann, the scientist's grandfather

We found information about him in the "service record of the doctor's assistant in the 1st *okrug* (district), titular counsellor Alexander Friedmann" who served in the Royal Medical Regiment of the Palace Administration.

The record contains an official list similar to a contemporary personal registration form. Information in the list was given not horizontally, but vertically – in columns. In column 1 was specified the civil rank, position, age, religion and salary; in column 2, the estate to which the parents belonged; in column 7, where the subject was educated and did his service; in column 14, the marital status, etc.

So Alexander Alexandrovich Friedmann was born on February 7 (Old Style), 1839. The place of birth is not indicated, the data about parents were not required in official lists – and these questions remain unanswered. But since A. I. Friedmann writes briefly that he comes from "military cantonists," one can say with certitude that he was born somewhere in the Western part of Russia into a Jewish family, and that in compliance with a ukase of Nicholas I on military cantonists (issued in 1827), he was recruited into the cantonist school, where there were Jewish teenagers among the trainees. Conditions in these schools were extremely harsh – much was written about this in the last century. They were harsh even compared with the service of recruits in the Russian Army of the time, which is known from classical Russian literature. But the extremely hard life of Jewish cantonists could become merely hard as soon as they adopted Orthodoxy. Alexander Friedmann also seems to have done this while still at the cantonist school, then in 1855 he moved from the abolished Revel semi-battalion of military cantonists (Revel is the old name of Tallinn) to the Revel Military Hospital to train as a doctor's assistant. Thus he got involved in medicine. From the Revel Military Hospital he moved to the Riga Military Hospital, and then in 1856 to the doctor's assistants' training school of the Second St. Petersburg Army Hospital. There he was promoted from junior doctor's assistant, to doctor's assistant, to senior doctor's assistant; and in 1873 he got a position as medical assistant in the First Court Medical Unit where he worked until he resigned in 1907 after "34 years of irreproachable service." During that time he won several (lower) orders and medals, the civil rank of *gubernsky* (provincial) councillor in 1887, and the rank of *kollezhsky* (collegiate) councillor in 1902. Friedmann's job was to provide medical treatment to the families of officers in the company of grenadiers guarding the Tsar's palace.

We learn further the name of A. A. Friedmann's paternal grandmother. Her name was Elizaveta Nikolaevna and she was a soldier's daughter, who was born on August 22, 1848, and died in 1907. The Friedmanns had three children: Alexander Alexandrovich, the scientist's father, Leonid Alexandrovich (who lived with his large family early in this century in Veliky Ustyug, worked in private service, and received 50 rubles a month), and a daughter, Maria Alexandrovna, who worked in a town hall (receiving 50 rubles a month) and gave music lessons. The information about their pay was important, because depending on it their father Alexander Friedmann was entitled to a certain pension upon retirement.

Alexander Ivanovich Friedmann died on October 14, 1910. One can find many interesting details studying his file and other records of the Court Medical Unit in which his name is mentioned. These details provide a picture of the time and its passions; past joys and sorrows. Let us only note one thing. When his wife died, A. I. Friedmann married

Lyubov Yakovlevna Anufrieva, a midwife, who was 25 years his junior. He adopted her son and then Alexander Ivanovich Friedmann had two sons, both called Alexander Alexandrovich Friedmann – a rare case indeed!

Ignaty Kasparovich Voyachek, the scientist's maternal grandfather

The most detailed materials about I. K. Voyachek are in his file kept in Collection 497 at the Central State Historical Archives. Before summarizing the materials of this collection let us quote the *Encyclopedic Dictionary* of Brockhaus and Efron: "Ignaty Kasparovich Voyachek, a specialist in the theory of music, was born on December 4, 1825, in the town of Zlín, in Moravia. Voyachek devoted his work to the study of Czech folk music and collected many folk songs and legends and dances. In 1857, Voyachek came to St. Petersburg and became a military band master. Later he joined the orchestra of the Mikhailovsky Theater. Voyachek was for some time an organist in the Italian Opera. In 1862, he was invited to the St. Petersburg Conservatory to teach elementary theory of music, and in 1866, he was appointed Professor of Orchestration. At present [1892 – V. F.] Voyachek is an organist in the Russian Opera, where for some time he served as second conductor." Riemann's dictionary provides additional information – Ignaty Voyachek graduated from Vienna University, was taught music by his father, worked as a bandmaster to the household troops of the Preobrazhensky Regiment, was invited to the St. Petersburg Conservatory by Anton Rubinstein, and wrote piano pieces which were published.

Additional information about Voyachek can be found in both above-mentioned archives (CSHA and SHAL), containing his personal files. He entered the state service in May 1857[2] as a bassoonist in the orchestra of the opera house, and he did not get a salary, but was paid per performance (three rubles for one performance, which would make a large sum if paid daily). In 1859, Ignaty Voyachek became a staff musician, in 1864 he became an organist, and in 1869 he replaced E. F. Napravnik as choirmaster of the Russian Opera. From 1876, he was conducting the orchestra, receiving 25 rubles per performance. From the files one learns the

---

[2] Voyachek came to Russia earlier: on August 17, 1855, he was given a travel certificate which contained his verbal portrait and which is kept in his CSHA file. The "portrait" described him as thirty years of age, of medium height, fair-haired, with grey eyes, a middle-sized nose and elongated face. Voyachek crossed the empire's borders on his way to the capital in a horse-drawn carriage, and station masters were ordered to provide him all possible assistance.

name of Friedmann's second grandmother: at the time of her marriage she was a colonel's daughter née Olga Ivanovna Meller. Furthermore, we found the name and the patronymic of Friedmann's great-grandfather: Ivan Karlovich Meller, who was apparently German, so Friedmann came from a mixed background. Ignaty Voyachek and his wife Olga had two children – a daughter Lyudmila (born on September 7, 1869) and a son Vladimir (born on December 7, 1876) already familiar to us.

Voyachek published a collection of military marches and by 1901 had become a professor extraordinary at the conservatory. He died (according to the *Musical Encyclopedia*) on January 28 (February 10) 1916, in Petrograd – at the age, as one can easily calculate, of 90.

Alexander Alexandrovich Friedmann, the scientist's father

We found detailed information about him in an extensive file kept in the CSHA, entitled "About the service of the ballet dancer Alexander Friedmann." It was opened on May 19, 1882 and closed exactly twenty years later, when at the age of 36 "the leading dancer of the ballet troupe" applied for and obtained a retirement pension as the required term of service in the ballet was over. In between were recorded the major events of his life. We learn that Alexander Friedmann was born on May 19, 1866, was baptized on May 29, finished at the St. Petersburg Theater School on May 27, 1884, and was soon admitted to the corps de ballet.

Another important document directly related to the main subject of this book is the official request of the dancer Friedmann to "His Excellency the Director of the Ballet Troupe" Alexander Frolov: "Wishing to enter a first marriage with the daughter of Professor Ignaty Kasparovich Voyachek, Lyudmila, I beg leave to request Your Excellency for a certificate in this matter." This request was submitted in October 1885, with the certificate issued that "the dancer of the corps de ballet Alexander Friedmann has been in the service of the theater's administration since May 19, 1882 and is registered as a bachelor."

A certificate on graduation from the St. Petersburg Conservatory is attached to the file, from which one sees that Friedmann studied there, as we would say now, without giving up his work in the ballet. His teachers in composition were Professors Johannsen and Rimsky-Korsakov (classes of canon and fugue). He took an advanced course in piano playing with fairly high achievements and completed his higher musical education in 1891, having passed all the examinations, including the history of music.

In the papers of A. A. Friedmann, Sr. there are two documents telling us about his work as a composer. He wrote music for two one-act ballets – *The Boatmen's Holiday* and *Amour's Pranks* – and each time (in 1890 and 1894) he had to put in a special application for remuneration for his work, valued each time at 300 rubles, which was about half the annual salary of a corps-de-ballet dancer.

In 1897, Friedmann, while continuing his work in the theater, began work as a bandmaster in the household troops of the Preobrazhensky Regiment. He seems to have owed this position to the recommendation of his father-in-law Ignaty Kasparovich Voyachek, who, as we know, worked in the same field. Friedmann's new and main service, to which he moved upon retirement from the theater, brought him minor civil ranks, orders and medals, and travels around Russia and abroad with the orchestra he conducted.

Friedmann's file has also a sad entry. In 1896 the Consistory at his request dissolved his marriage with Lyudmila Friedmann. The official certificate says that the church sided with A. Friedmann, the plaintiff. It is known that his 9-year-old son stayed with his father. He did not see his mother until 1920 (there could only have been short encounters), living either with his aunt, Maria Alexandrovna, or with the father and his new wife, Anna Khristianovna Ioganson.

Lyudmila Ignatievna Voyachek, the scientist's mother

Very little is known about Lyudmila Ignatievna who, as has been said, was born on September 7, 1869. In 1881, her father, Ignaty Voyachek, placed her in a school attached to the St. Petersburg Conservatory "to study piano, and other scientific subjects." She had hardly reached 16 when she married Friedmann's father. As we already know, theirs was an unhappy marriage. Valentina Vitalievna Doinikova, whom we shall meet again later, recalled that in the 1910s Friedmann's mother expressed a wish to meet her son, but he rejected her suggestion. Their meeting seems to have taken place when Friedmann, after several years of absence, returned to Petrograd. In one of Friedmann's papers dating from the first half of the 1920s, it is said that his family consisted of his wife, his mother-in-law and his mother, and that all of them were living together in the Fifth Linia on Vasilievsky Island.

Let us go back to the above-mentioned conversation of one of the authors of this book with V. I. Voyachek, substituting the more informal first person for the academic third person.

I (V. F.) went to visit Vladimir Voyachek in the summer of 1971. The weather was hot, but when I was shown into his study, I saw a very old man in a general's uniform, wrapped up in a rug. I was struck by the cautious way he began to inquire why I was interested in Friedmann and why I came to visit him. He had problems with hearing. I had almost to shout that Friedmann was a great physicist, the author of a theory which excited the whole world and which described the evolution of the Universe, and that I came to him, Voyachek, because he was one of Friedmann's few remaining relatives. "Don't you know that he has a son?" Voyachek asked. "Yes, I do, but he was born after his father's death and he will hardly be able to tell me anything." "What are you interested in?" I listed several questions, more or less standard, and added that I was interested in everything that he could recollect about his nephew.

He said, "Please type your questions on a typewriter, in capital letters, and once again tell me about yourself – where do you work? And indicate where articles about Friedmann have been published. I would like to read them. Send me all this and ring me up."

We agreed that I would do that. The same evening, on August 3, I typed three pages with questions and the information which Vladimir Voyachek asked for and sent this to 3 Klinichesky Street, where he lived in one of the blocks belonging to the Military Medical Academy, and rang him up later. On the appointed day and hour I was again at his place.

Vladimir Ignatievich told me very little. His sister, Friedmann's mother, was a pianist, a conservatory graduate, and taught music in her last years. She died in 1953, at the age of 83. Friedmann's father graduated from a ballet school (do you know, Voyachek asked, that it is on Rossi Street?). "My father-in-law," Voyachek went on, "was Viktor Lvovich Kirpichev. Is this name familiar to you?" "Certainly! He was a distinguished researcher in mechanics." "As a matter of fact, it was through Kirpichev that Friedmann got acquainted with professors at the Polytechnical Institute. If you're interested, I can tell you that Kirpichev was Witte's consultant in the organization of polytechnical institutes – in Kiev, Kharkov and St. Petersburg."

"And do you know," Voyachek went on, "that Friedmann got involved in aviation? Aviation was recruiting scientists at that time and the clinic of the Medical Academy was among the first to develop avaiation medicine. He used to rush wildly into all sorts of adventures."

Here Voyachek brightened up: "He was like a star which lit up and disappeared. He and I did not even manage . . . [there is a gap in my notes here. – V. F.]. We began to see each other in 1920. The later years went by like a flash. But I will always have a feeling of loss. This loss of my relative made me feel destitute. Everything used to go along so well with him. Fate strikes people without any guilt on their part! He was coming back from a southern resort, drank some unboiled water, fell ill . . ."

Vladimir Ignatievich became silent. It was clear that it was hard for him to speak. I rose to leave. "Are you leaving? Would you like to come this way, please." He showed me to the hallway. And he said when parting: "Young man! [I was already over 40 – V. F.] If only you knew how lonely I am! All my relatives, my friends, my students are no more. I feel as if I am on an alien planet!"

I felt terribly embarrassed that I had upset such an old man, and I hurried to say good-bye, making up my mind never to disturb him again.

Now, over 20 years later, I see that V. I. Voyachek was accurate in his answers to those few questions that he selected from the ones I asked. I understood his caution too: it appears he did not want to touch upon the family drama, one of whose characters was his sister.

I was struck at that time, as I am now, how often the interest in outstanding people is so much belated. The scope of Friedmann's talent was already evident in his lifetime. It was even more highly appreciated with the advance of astrophysical theories which began to be developed in the second half of the 1940s. Friedmann's mother was still alive at that time; his uncle was not yet too old; one could find his relatives in the father's line; dozens of his students and colleagues were alive and fully active. Among them were Friedmann's school and university companions, Academician Vladimir Ivanovich Smirnov and Professor Alexander Felixovich Gavrilov. My father, Ya. I. Frenkel, worked with Friedmann for more than four years at the Faculty of Physics and Mechanics in the Polytechnical Institute. His small home library had, since pre-war years, two books by Friedmann: one, written jointly with V. K. Frederiks, was a mathematical introduction to a course in the theory of relativity. The book came out in 1924, and its title page had the author's dedicatory inscription. The other book, *The Hydromechanics of a Compressible Fluid*, was published posthumously in 1934. The book opened with Friedmann's portrait, the same which opens the present book. The sad expression of his face in this picture seemed natural to me when my father told me about his untimely death. He added that he was an exceptionally gifted scientist, but – alas! – I did not ask him to tell me more about Friedmann at the time. As a matter of fact, among the relatively few people whom I met, many were Friedmann's close acquaintances – Professor R. O. Kuzmin, and researchers at the main

Geophysical Observatory named after A. I. Voeikov. I was once offered an introduction to a person who served in the army with Friedmann, near Przemyśl, and I kept postponing this meeting until the man died. Only Valentina Vitalievna Doinikova, on her own initiative, told me something about Friedmann, generously sharing her reminiscences about P. S. Ehrenfest. This brings to mind what was said by an anonymous French wit in the middle of the last century: "No matter what we say every day and no matter what we do every day, with each passing day there are fewer people among us who knew Napoleon personally." How few people are left among us who personally knew Friedmann! What precious testimonies about his life are irretrievably lost! . . .

2

At the 2nd St. Petersburg Gymnasium

At the turn of the 20th century, St. Petersburg had 15 secondary educational establishments – classical, modern and military schools. Among the teachers of physics and mathematics we see names which became famous far beyond the history of secondary education. In the 4th Gymnasium, physics was taught by Apollon Pavlovich Afanasiev, who later became a university professor; another university professor, Karl Karlovich Baumgardt, taught physics and cosmology at the 8th Gymnasium named for Karl Mai, together with F. N. Indrikson, the author of the school physics textbook, who presented in his lectures a famous university general physics course of Professor O. D. Khvolson. Why did Friedmann's parents choose the 2nd Gymnasium as the place for Friedmann, Jr. to study for nine years? There appear to be two reasons. The first is that the gymnasium was relatively close to the house where they were living at the time. Friedmann's father occupied Apt. 4 in 35 Moika Street. The house exists today, under the same number; furthermore, as the cast-iron plaque on it says, it is protected by the state as an architectural monument of the early 19th century. Its corner looks onto the Zimnyaya Kanavka (canal, or, literally, ditch) and is next to the Naval Archives building, started in 1883 and completed one year before Friedmann's birth – in 1887; an inscription to that effect appears on the impressive facade looking onto Khalturin Street, formerly Millionnaya Street. If one comes out from under the arch, above which there are the second and third storeys of the house, decorated with columns, then on the right is No. 37, whose corner looks out onto Palace Square, on the left is the entrance to the Chapel, and a little further left, along the canal, on its opposite bank one can see the front of No. 12 Moika Street – Alexander Pushkin's last address. This is an old part of the city without new residential districts. If one comes out onto Palace Square, crosses it and, passing Nevsky Prospect, moves along Admiralty Prospect to Isakievsky Square, then right behind the Mariinsky Palace (now the building of the City Soviet's Executive Committee) is Grivtsov Lane (formerly, Demidov Lane), and behind it is Secondary School 232 – the former 2nd St. Petersburg Gymnasium. It is the preparatory class of this school that Alexander Friedmann began to attend in 1897, first, of course, accompanied by one of his relatives – his father, grandmother or aunt (his grandfather, Alexander Ivanovich Friedmann, was living very close to 35 Moika Street, in a beautiful residential block on Dvortsovaya Embankment, at No. 32, which is today decorated by two memorial plaques certifying that Giacomo Quarenghi used to live there in the past, and in our times Academician I. A. Orbeli). It seems that the very route from Alexander's home to the school (it could vary too: one could reach Nevsky Prospect through Moika Street and turn into Bolshaya or Malaya Morskaya Street – today Herzen Street and Gogol Street respectively) was in itself a tool of aesthetic education – first of the boy and then of the youth: so beautiful are the embankments, streets, avenues, squares, residential houses, monuments and palaces which he passed every day!

The 2nd Gymnasium, the oldest secondary educational establishment in St. Petersburg, was founded in 1806, i.e. close to the time about which Alexander Pushkin said: "Of Alexander's days the beautiful beginning." In its centenary year, just as Friedmann was in the eighth and final class, it was again renamed and became known as the Emperor Alexander I Gymnasium, "by reason of the fulfilment on September 7 of one hundred years from the day of its foundation by the wish of Emperor Alexander Pavlovich, who died in the Lord," as the relevant ukase expressed it.

The Director of the Gymnasium in Friedmann's time was Alexei Ivanovich Davidenkov, Actual Councillor of State, the chairman of the Society for the Relief of Needy Students. The name Davidenkov is well known in the USSR, particularly in St. Petersburg. Nikolai Nikolaevich Davidenkov, a Full Member of the Ukrainian Academy of Sciences, was a famous researcher in mechanics and physics. His brother, Sergei Nikolaevich, was a full member of the Academy of Medical Sciences, and a major psychiatrist. The Director of the 2nd Gymnasium was their uncle. He was born in 1853, and upon graduation from the History and Philology Faculty of St. Petersburg University he taught first in gymnasia in Riga, and from 1876 in St. Petersburg. Davidenkov became Director of the 2nd Gymnasium in 1898, and judging by the minutes of the sessions of the Pedagogical Council, was a progressively-minded person, although he would not enter into conflicts with the Ministry of Education which directed the work of the gymnasia.

The files of the 2nd Gymnasium in the Pskovsky Street archives are in perfect order and make it possible to understand and get a feeling for what the life of this school was like – a good chapter for a book on pre-revolutionary secondary schools in St. Petersburg! One has to start somewhere . . . Here is Collection 174, File 3972, List 1 – "the register of progress and conduct of the students of the school" for 1898, when our hero was in the preparatory form. On page 3 of the register there is an entry on Alexander Friedmann, the son of an artiste, of Orthodox faith, born on June 4, 1888, entered the school in August 1897. There are entries concerning all four terms and such subjects as Russian, catechism, mathematics, calligraphy and drawing.

When one comes across such data about a man who later became an outstanding scientist, it is of interest, irrespective of plus or minus. If the person got good marks, it was only natural! If he got bad marks, there is nothing surprising either – it is not the marks (particularly in the first form!) that are important, what matters is talent, which in the first place cannot be regimented, and in the second place does not manifest itself at once. And yet, one can only be surprised by the abundance of merely satisfactory marks in Friedmann's register! It is only in catechism that the boy had mainly excellent marks: in all the terms and in all the aspects (attention, diligence, achievement) he had nine excellent marks and only three good marks – and the latter are solely in the first term. There are no excellent marks in mathematics, mainly satisfactory marks; the same goes for the Russian language. In calligraphy, with the same number of satisfactory marks, instead of good marks he had bad marks. The "drawing" column is adorned with 12 "threes"[1] – here Friedmann was consistent. Today we may smile looking at these marks, but the boy must have had a lot of troubles because of them and may have been upset.

There were about 40 students in the preparatory class, all from non-gentry strata – children of civil servants (not higher than councillors of state), clergymen, medical doctors, merchants, other middle-class town dwellers, peasants – but none from working-class families! The same pattern was maintained in later years, with only a few children of the nobility.

What was a regular day of these nine- or ten-year-old boys like? They came to school by 9 o'clock. Classes lasted for 50 minutes; the first break

[1] "Five" stands for excellent, "four" stands for good, "three" stands for satisfactory, "two" stands for unsatisfactory, and "one" stands for very bad.

was 10 minutes, the second 20, the third (the longest) 35, the last one again 10 minutes. Only in the final two years did they have six classes a day.

Now, jumping several years ahead, let us look at the 1901–02 school year. By now Friedmann was in Class 4, section I (classes were distinguished not by letters, as is the custom today – class 4A, 4B – but by numbers: Class 4-I, Class 4-II). Friedmann had almost solely "fives" in mathematics, Russian and Latin. There was another student in Class 4-I, whose achievements were as high. His name was Yakov Tamarkin; for many years his name and Friedmann's were to become inseparable, like Ilf and Petrov.[2]

Let us look at the minutes of the Pedagogical Council for 1903. Friedmann was in the fifth form. The composition of the class had changed a little; some had dropped out, some new students had come. There were 35 students, of whom, as is seen from the list, 20 were the children of the nobility and of civil servants, 9 were from the urban middle-class families, and 6 were from peasants' families. On completion of the fifth form, five students were given awards of the first and second degree (similar to today's honor certificates). Friedmann and Tamarkin got a first-degree award – and this was to happen every year.

Later, in the final forms, in addition to the general evaluation of his progress (of how he prepared his lessons, how attentive he was, whether he displayed interest in his studies and, of course, whether he played truant) it is recorded of Friedmann: "He studied mathematics on his own and was keenly interested in the subject." He got "fives" in all subjects.

And what about the conduct of the student A. Friedmann? If you have read Lev Kassil's book *Konduit and Shvambrania*, you will know that the answer to this question is to be found in a *konduit*.[3] These are kept in archives in perfect order. I have before me a document which has never been asked for in over eighty years – File 4323 (Collection 174, List 1) – "the *konduit* register for the 1904–5 school year." The register was made in a printing shop, its pages were divided into columns: "month and data," "student's name," "act of misconduct," "teacher's name," "punishment." The *konduit* has 79 sheets, i.e. almost 160 pages, with about ten entries on each page – 1600 entries in all! I am looking through the register page by page. And there is no Friedmann yet! I see the names of the rascals: e.g. A. Pogorzhalsky "does not stay at his desk after the bell has rung for the class to start," "when walking to have breakfast stepped out of line" – an "out-standing" boy! – "failed to submit the week's

[2] I. Ilf and Ye. Petrov are famous Soviet authors who wrote books jointly.
[3] Register of school students' conduct.

progress note with his parents' signature" – seemed to be afraid of his parents' anger, "brought a rattle to the classroom and played with it before the teacher came in" – that is all on one page. As a matter of fact, such acts of misbehavior were not too severely punished – a student was made to stay in the classroom for one, two or three hours after classes. The punishments were not generally severe (and the "crimes" were not dangerous) – one is surprised sometimes by the teachers' patience. Could this be due to the influence of the director, A. I. Davidenkov, and his calm way of directing the school, about which we will have a chance to speak below?

A. Friedmann was not to be found in the conduct register, but – as a consolation prize – Yakov Tamarkin appeared at the very end. On November 4, 1904, he "behaved outrageously before the class, and before the gymnastics class he romped and shouted," for which the teacher N. Kuznetsov made him stay in the classroom for an extra two hours. One more thing – Tamarkin was often late for school.

However, for the students the *konduit* was a very serious thing. If after the gymnasium they entered an institute of higher learning, a special page was attached to their file – "the *konduit* of the conduct of the student so-and-so of the gymnasium such-and-such." The entries from the *konduit* for the last two years of studies were transferred onto this page.

The teachers

Among the teachers of the 2nd Gymnasium, we are interested most of all, of course, in mathematicians and physicists. Mathematics was taught by Yakov Varfolomeyevich Iodynsky and Pyotr Nikolaevich Hensel. Both graduated from the Faculty of Physics and Mathematics of St. Petersburg University at about the same time (the mid 1890s). Hensel got a first-degree diploma, and Iodynsky a second-degree diploma (the latter corres-ponded to "satisfactory achievements" in examinations). Both mathematics teachers seemed to be aware of how talented some of their students at the school were – suffice it to mention Vladimir Ivanovich Smirnov, Friedmann and Tamarkin (who were one year younger than Smirnov) and another brilliant young fellow of whom we will speak later – Mikhail Petelin. There are some documents which provide evidence of Iodynsky's support for Friedmann and his schoolmates – they will be cited later.

Physics and cosmography were taught by Ivan Vasilievich Glinka, who had also graduated (in 1901) from the Physics and Mathematics Faculty of the University (with a first-degree diploma). He was highly esteemed by

Professor S. P. Glazenap, a famous astronomer, who had recommended him to Davidenkov – it seems that to get into this secondary school was not easy. Glinka was loved and respected by the students, not only because he was well-versed in the subjects which he taught in the school, but also for his activities outside his duties. Glinka's personal file (col. 174, list 1, f. 4918) contains a letter from the superintendent of the St. Petersburg Educational District to the Director of the 2nd Gymnasium, A. I. Davidenkov, dated January 10, 1906, which says: "A teacher at the school you are in charge of, I. V. Glinka, wrote in December 1905 a statement on his joining the political strike. This statement resulted in Mr. Glinka's being invited to give an explanation, in the course of which he recognized the correctness of the principle that there should be no political struggle in the school, as it has a harmful influence on the students, and confirmed the latter statement in writing. Now, I. V. Glinka has again submitted a statement in which he is asking for his previous statement to be considered invalid" (page 18). The letter further demands that Glinka be called to order. Towards the end of this chapter we shall outline the political climate in the Gymnasium. Against that background, Glinka's support of the school students' movement, the evidence for which is provided by the above quotation, was particularly important.

Glinka's educational activity was probably even more significant. In 1903, he organized in the Gymnasium optional classes in physics with students of the eighth (final) forms. The classes were held once a week in the physics classroom. By the time Friedmann reached the eighth form, i.e. in 1905, a physics circle had been organized which was attended by all senior school students who wished, three times a week. The most active members of the circle, the monitors, supervised work in the laboratory. They were elected by the members of the circle, which soon got the solemn name of "the Society of Lovers of Physics of the 2nd St. Petersburg Gymnasium." At sessions of the circle reports were also given. A library was also set up which subscribed not only to physics books, but also to popular science ("elementary" as they were called at the time) magazines, including the famous *Bulletin of Experimental Physics*. Moreover, the students started their own journal, in which, besides the accounts of practical classes and talks given at seminars, there were overview articles by the circle members; in these, the authors attempted to explain the phenomena which they dealt with in their oral presentations. The students also published their "proceedings," or "the Society's News Bulletin" – here one can easily assume the influence of Friedmann and Tamarkin (although there is no direct evidence of their participation in the work of the circle), because, firstly, by this time the two youths already knew well the published proceedings and "news bulletins" of various learned societies and, secondly, they were actively involved in similar activities at the University a year or two later. It is interesting to note that, according to N. N. Andreyev's memoirs, Friedmann frequented his school, the physics classroom specifically, after he had become a university student.

The limits of this book do not, unfortunately, allow us to give information about the sort of textbooks which were used in teaching algebra, geometry, cosmography and physics, and to relate their contents to the most recent achievements of physics of that time. Through the courtesy of the teachers of School 232 and V. B. Belodubrovsky we got hold of the "notebooks" of Georgy Karayev, a student of the 2nd Gymnasium who was four years younger than Friedmann. The notebooks (school reports) are kept in the school museum. These notes dating back to 1905–06 (the fourth form) and the next (fifth form) school year show how little attention was given in these forms to mathematics: only five periods a week in the fourth form, and six periods in the fifth! And it is in the fifth form that Friedmann and Tamarkin began to study mathematics really hard![4]

Examinations

The documents kept in the archives show us what kind of mathematics problems were offered to Friedmann and Tamarkin at the oral and written matriculation examination in the spring of 1906. They give some idea about the level of teaching and the requirements the students were to meet, and we think it useful to provide them here. Here is the program of oral examinations with an algebra problem set by Ya. V. Iodynsky.

Determine the 16th term in the expansion of $[\sqrt[10]{(1/c)} - d]^m$, where

(a) m equals the root of the equation $x - \sqrt{(245 - x)} = 5$;
(b) c equals the largest of the values of y which are obtained by solving in positive integers the equation $35x + 66y = 2718$;
(c) d equals the number whose logarithm to base $\sqrt[8]{2}$ equals $-\frac{4}{3}$.

Problems in geometry and arithmetic were set by P. N. Hensel.

Geometry. One of the parallel sides of a trapezium is 7.284 ft, the other side is 5.328 ft; one of their non-parallel sides is 3.786 ft. The angle between the latter side and the longer base is the root of the equation $4\cos 2x + 3\cos x = 1$. Determine the angles and the fourth side of the trapezium.

[4] From Karayev's school report one can learn a lot of interesting things – thus, there were no marks put in the daily report as is done today in the USSR.

Arithmetic. Three merchants decided to form a partnership and contributed different capitals. The first merchant contributed a capital which would turn into 21,490 rubles if it were deposited for 0.583 years at 4% interest. The capital of the second merchant, being deposited for 0.61 years at 5.09%, would give 450 rubles profit. The third merchant's capital was $\frac{5}{6}$ that of the second. After one year of trade the merchants made a profit which was equal to the least common multiple of the numbers 270, 300, 405. How many rubles out of this profit should each of them get, and what percentage of their original investments does their profit represent?

The written tests in algebra and geometry were prepared by Iodynsky.

Algebra. Solve the indeterminate equation $ax + by = c$ in positive integers, where a is equal to the root of the equation

$$\frac{x-1}{1+\sqrt{x}} = 4 - \frac{1-\sqrt{x}}{2},$$

reduced by 5, b equals the fifth term of the geometrical progression in which all terms are positive, the second term is greater than the first by $1\frac{1}{3}$, and the difference between the fourth and the first term is $14\frac{4}{9}$, and, finally, c is equal to the coefficient of the term in the expansion of

$$[z\sqrt[4]{(z^3)} + \sqrt[8]{(1/z)}]^m$$

which contains, after simplification, the fifth degree of z and where m is equal to the root of the equation

$$\frac{\log x}{2 - \log 5} = 1.$$

Geometry. In a semicircle of radius R there is a chord CD parallel to the diameter AB and spanning an arc a; from D a perpendicular DE is dropped on AB and E is joined to C. The figure limited by the straight lines DE, EC and the arc DC evolves about the diameter. Determine the volume of revolution if it is known that $R = 23.476$ m, and the arc a is determined by the equation $5\cos 2a + 13 = 24\cos a$.

One would guess that Friedmann and Tamarkin, who had had their article published in a German mathematical journal, had no difficulty in passing the examinations. This is evidenced by the matriculation certificates which are attached to the personal files of the students of the Physics and Mathematics Faculty of the University. This is what Friedmann's certificate looks like:

The Emperor Alexander I Gymnasium
(former 2nd St. Petersburg Gymnasium)
MATRICULATION CERTIFICATE

This is given to the *gubernsky* councillor's son Alexander Friedmann, of Orthodox faith, born in the city of St. Petersburg on June 4, 1888, who has studied nine years in the Alexander I Gymnasium and spent one year in the eighth form, to certify that, firstly, on the basis of observation during his studies at the 2nd St. Petersburg gymnasium his general conduct was excellent, his attendance was good and his written papers were satisfactory. He displayed sufficient diligence and particular interest in the study of mathematics.

And, secondly, it was found . . .

Then there are listed the marks given by the Pedagogical Council and those based on the results of examinations. Further, there are listed the examinations which Friedmann took from April 12 until May 31(!). They are catechism,[5] Russian,[5] logic,[5] Latin,[5] Greek, mathematics,[5] physics, mathematical geography, history,[5] geography, French,[5] German.[5] In all these subjects Friedmann received excellent marks. This is how the certificate ends:

In view of his consistently excellent conduct and diligence and excellent achievements in science, particularly in mathematics, the Pedagogical Council has decided to award him a gold medal.

The certificate is signed by Director A. I. Davidenkov and the teachers.

By the time Friedmann graduated from the school there were 32 students in his class; four of them got silver medals and three (Tamarkin, Friedmann and V. Engelke) gold medals (in Class 8-II there was only one student who was given a silver medal). All in all, about 65 people finished the gymnasium in 1906. All of them passed the examinations. The records of the Pedagogical Council have information about the way the young people continued their studies. Since this is of a certain "social" interest, let us give the respective figures: 49 graduates entered universities (the Faculty of Oriental Languages 2, the History and Philology Faculty 3, the Physics and Mathematics Faculty 9; the Faculty of Law 16, the Medical Faculty 19); 13 entered specialized higher schools (the Technological Institute 1, the Institute of Railway Engineering 4, the Polytechnical Institute 8); finally, 3 chose military service and entered military schools.

The records of the Pedagogical Council have an interesting "prophetic" entry: "Of those who have completed the course, the most promising, as to further progress in mathematical sciences, are Tamarkin, Friedmann, Kapustin, Staropolsky and Ivanov." Well, as a matter of fact, this prophecy was fulfilled in at least 40% of the cases. The teachers of Friedmann and Tamarkin had good grounds for this prediction. They knew about the above-mentioned article by the two school students on

[5] These subjects were evaluated both by the Pedagogical Council and at examinations.

Bernoulli numbers, published in a German mathematical journal, and about Hilbert's review of the article. They also knew that the two teenagers had studied a large number of mathematical treatises and textbooks. In one of his reports kept in his university file, Friedmann gives those titles. Quite an impressive list it is! Let us give some of the titles from the list: Chebyshev's *Theory of Congruences*, Sokhotsky's *Higher Algebra*, Euler's *Elements of Algebra* in two parts (in French); works by Dirichlet, Dedekind, Legendre, Markov, Zolotarev, Bachmann and Hilbert in the theory of numbers; works by Sokhotsky and Weber in higher algebra; works by Euler, Goursat, Bertrand, Serret, Stolz, Cauchy, Picard and Jordan in analysis; works by Bobylev, Appell, Riemann & Weber, Poincaré and Clausius in mechanics and mathematical physics; works by Andreyev, Bukreyev, Bianchi and Lamé in geometry.

Friedmann goes on to write (having Tamarkin and himself in mind): "Unfortunately, there was a serious gap in our knowledge of non-Euclidean geometry and synthetic geometry, which, we hope to fill in our further studies (at the University – V. F.). Towards the end of our studies at the Gymnasium and at the beginning of our studies at the University we gave attention to the foundations of our science. We have studied the following works . . ." Then follow works by Schoenflies, Du Bois-Reymond, Borel, Dedekind and Cantor. Friedmann notes in conclusion that they thoroughly studied works by Markov and Bunyakovsky in the theory of probability and those by Markov in the calculus of finite differences.

No wonder that when applying to the University, in addition to his honors school diploma, Friedmann submitted the following "Certificate" typed on the school's official form:

A. A. Friedmann received the matriculation certificate in 1907. From the third form he independently studied elementary mathematics; from the fifth form he studied physics and higher mathematics under the supervision of Academician Markov. He concerned himself with the following areas of higher mathematics:

theory of Bernoulli numbers;
theory of prime numbers;
theory of congruences and theory of elementary functions.

In the theory of Bernoulli numbers he pursued the question of some formulae of permissibility of Bernoulli numbers, developed some general forms comprising prime numbers, and found a formula determining the number of prime numbers less than a given one.

In the theory of congruences he gave a formula expressing the number of congruences in general form, and developed in detail the solution of congruences of the second degree; an article on this question was accepted by Hilbert to appear in the mathematical journal *Math. Annalen* published in Leipzig.

At the final examinations on May 24, 1906, Mr. Friedmann presented the theory of congruences and the theory of trigonometric functions. I consider it my duty to certify this on the basis of the report submitted by the mathematics teacher Ya. V. Iodynsky and the physics teacher I. V. Glinka.

The certificate was signed by A. I. Davidenkov.

This certificate should be given a short commentary. Friedmann's first wife Ekaterina recalled his story that when Tamarkin and he learned that their article had been accepted for publication they got so excited that they were expelled from the classroom. However, it seems that after learning the reason for their misbehavior, the teacher did not record it in the conduct register (*konduit*). Another commentary concerns the acquaintance of the school students with A. A. Markov. I. I. Markush, who studied Tamarkin's biography, reports that Tamarkin and Friedmann attended the sessions of the city's mathematics seminar for secondary school students and there they attracted the attention first of B. M. Koyalovich and then of the illustrious mathematicial A. A. Markov.

Political activity

Their passion for mathematics, studies in the school circle, writing joint papers and taking part in the sessions of the students' mathematical society should have taken up all Friedmann's and Tamarkin's free time. But as a matter of fact, they were also involved in very intensive social – or, despite the youths' age, let us use a stronger word – political activity. The evidence of such activity is reflected not only in occasional phrases found in memoirs about Friedmann, but also in documents and even publications – at least in one book published in 1926 and written by Friedmann's friend, a former student of the 10th St. Petersburg Gymnasium, S. Dianin. This book gives a fairly comprehensive picture of the activity of students at gymnasia, technical high schools, and district (*zemstvo*) schools; activity which, overflowing beyond school bounds, linked with the revolutionary movement that spread across Russia on the eve of the First Russian Revolution and reached its peak at the end of 1905. But let us first look at some materials from the archives of the 2nd Gymnasium.

The events which took place in the school in October were triggered by a tragic accident involving one of the students, Boris Kolyshkin, who committed suicide. This tragedy excited the whole of the school and is reflected in the materials of the Pedagogical Council, where it was given a detailed analysis (because one of the versions gave the severity of the teachers as one of the reasons for the suicide). The report about the student's funeral said that his schoolmates laid on his grave a wreath with the inscription "To the oppressed from the oppressed." Trying to justify themselves in their commentaries on this sad event, the speakers at the council session reminded (themselves first of all) that back in the spring of 1905 the school administration had abolished punishment by detention in a lock-up.

Prior to the events of October 17, 1905 (a general strike was planned for that day), there were already gatherings of senior students in the school. The students were particularly angered by the history teacher, P. K. Tikhomirov, whose classes were, in their opinion, conducted at a low level, completely ignoring the modern political and economic situation in Russia: the students demanded his removal. On October 11, a students' meeting adopted a resolution which was handed to the school's director through the janitor. This resolution is attached to the materials of the council and has thus survived. Let us quote its relevant part: "Taking into consideration that the presence of P. K. Tikhomirov cannot be tolerated; taking into consideration that only abnormal conditions in the school and throughout society made the appearance of people like Tikhomirov possible, the meeting of senior students of the 2nd Gymnasium demands a radical change in this condition. As a particular measure, the meeting demands the immediate dismissal of Tikhomirov from the school. Concerning the students of Class 5-I, whose fate is to be decided at the coming session of the Pedagogical Council, the meeting demands most resolutely that these students should not be punished in any way. Otherwise, the meeting will take most resolute measures." This is followed by a note saying that the school's director was allowed to be present at the meeting and tried to bring the students to "reason" in all possible ways, but the only visible success of his mission recorded in the materials of the Pedagogical Council was that the students eventually went home.

At the sessions of the Pedagogical Council held thereafter the decision was made to suspend classes (there are no entries in the above-mentioned conduct and achievement report from the 12th to the 24th of October, so it seems as if the school was really suspended). Nevertheless, the most radically-minded students broke into the school and held another meeting. On October 12 they adopted the following resolution (which is also kept in the materials of the council's minutes): "The meeting of senior students of the 2nd Gymnasium expresses its indignation over the closing of the school without prior discussion of this decision with the students. The meeting considers an untimely strike and closing of the school extremely harmful to the cause and expresses its resolve to join the general political strike, if it is to be held,[6] which is caused by the present developments."

What was our hero's role in all these events? Apart from the good reasons, which will become clear in what follows, for claiming that he took a direct part in drafting the above resolutions, the records of the Pedagogical Council contain an entry which touches on this question. It reads as follows: "On October 13, during the students' meeting, the father of A. Friedmann, a student of the 8th form, called out his son and, holding him by the hands, implored him to come back home saying: 'Mother is ill, let's go!' The student broke away from his father's hands and went back to the meeting saying: 'My fellow students are more important to me'." Let us emphasize the feeling of fellowship which was so characteristic of Alexander Friedmann. As to his father's argument, we join the younger Friedmann in our doubts as to the health of his step-mother. Reference to relatives' ill-health is a usual and not very original way of bringing youth to reason.

In October 1905, classes at the school resumed, then again were suspended several times, but the movement of the students continued and they succeeded in winning some rights, though limited. (Thus, for instance, students' meetings were now allowed, although only for senior students and in out-of-school hours). Let us quote one other resolution of a students' meeting which gives some idea of the mood of those distant years through the students' living words.

Outraged by the government's repressions applied recently to secondary schools, being aware that only the temporary reduction in the energy of the revolutionary movement could give the government the opportunity to take such measures, which basically cannot lead to anything and cannot suppress the secondary school movement, we, senior students of the 2nd Gymnasium,[7] having gathered at a meeting on January 23, 1906; (1) express our vigorous protest against the teachers' dismissal and our sympathy for them; (2) declare the positions of the dismissed teachers occupied by them and therefore declare them not available to anyone else; (3) concerning the possible expulsion of students from the school, we declare that if any students are expelled for their participation in a meeting or for a strike, there will be a strike until the students are re-admitted, and we shall assist in the holding of the strike by all possible means; (4) concerning the abolition of the institution of monitors, we express our protest against this and our firm wish to form an organization having a solely technical purpose of assisting in convening a students' meeting.

---

[6] The general strike planned for October 17 was declared on October 13.
[7] Note that in all the cited materials the school is called by its old name and not by its official name "Alexander I Gymnasium."

The meeting had tangible results: the council which held its meeting on January 24 did not dare to resort to severe repressive measures against the participants in the meeting. The minutes of this council's session have this: "Several members of the council said that at present it is impossible to accuse the students: seeing that their former school-mates enjoy complete freedom of assembly for meetings and rallies in higher educational establishments, and seeing that society is very much satisfied by their developments, they are unwillingly influenced by this atmosphere, and many of them are afraid most of all of being considered backward." The council decided to suspend classes in the school, but not to lock the door – to avoid excesses – and to admonish the students at the entrance. As is seen from what follows, the students were allowed into the school and had meetings and rallies during the events of October 1905, discussing both academic and political issues. The statement by the physics teacher Glinka was the most progressive in the debate at the council's session. From the records of these sessions we learn the names of the most disliked teachers, as well as peculiar forms of the students' opposition to the teachers they detested: "chemical obstruction," i.e. the use of calcium carbide having a strong smell, and "acoustic obstruction" (this is our own term, not borrowed, as is the previous one, from the minutes) – throwing crackers, singing songs and (again a quotation from the minutes) "other pranks."

The generally progressive policy of the school's Director A. I. Davidenkov has already been mentioned above. It was manifested in the diplomatic presentation of events in the minutes, which were, of course, monitored by the powers that be; these materials presented school students as immature children, scamps, who should be treated accordingly. We have seen already how ardent the students' meetings in the school were, how radically political issues and those concerning the students' life were formulated. How does this square with those parts of Friedmann's and Tamarkin's matriculation certificates and their schoolmates' certificates, which say that "during their school years their conduct was generally excellent"? The fact is that Davidenkov, at the council's session on April 24, 1906, i.e. prior to the students' assessment, raised the question of assessing the future university applicants' conduct in their matriculation certificates. It was noted in the minutes of this issue that "unrest on political grounds and participation in strikes, which have occurred in the current school year, go so far beyond normal school life that it is impossible to judge them by conventional standards, in view of which the council has decided to exclude from discussion all the events of the above-mentioned character and to assess the conduct of graduates only with regard to their everyday discipline."

Now let us turn to the above-mentioned book by S. Dianin. Its opening chapters present the prehistory of the movement of St. Petersburg secondary school students. Most important for us is Chapter 3: "The academic year 1905–1906." The printed text of this book fully supports the typewritten text of the above-cited minutes. It says that having returned to their gymnasia (and technical schools) after the strike of October 17, the students began their fight to change the established school rules for some form of control over the instructional process through the institution of elected monitors. In fact, in the students' movement there was a certain differentiation – the majority supported the Social Democrats, but there were also Socialist Revolutionaries and anarchists. (The "Anarchist" wing broke windows in gymnasia and even exploded a bomb in one of them). About half the city's gymnasia were particularly active, among them the 2nd and 10th, the latter being the one where Dianin was a student. The preserved materials contained in his book, especially the resolutions of students' meetings, show that some kind of common policy was worked out by the most active participants in the movement; a simple comparison of the texts of resolutions prompts this thought. Here is an excerpt from one such resolution – that of the meeting of the students of the 10th Gymnasium, which was held on October 25, 1905: "We consider it a crime against the people and physically impossible for ourselves to follow in our studies the guidelines provided by the people's enemy, the government which gave the October Constitution and organized a whole series of killings on October 18, for which reason we declare a strike." The book says that city students united into the "Northern Union." And it was at the meeting of students of the 2nd Gymnasium that it was proposed to hold an all-city council of all the elected monitors of gymnasia and technical schools. By December 1905 the "Northern Social-Democratic Organization of Secondary Schools" was formed, which was headed by a committee consisting of 15 people representing at first about 100 (in April, 200) students. This committee, which was soon named the Central Committee, consisted mainly of representatives of St. Petersburg secondary schools, which were divided into four districts: the Vyborg district, the Vasilievsky Island district, the City district (which included the 2nd Gymnasium, as well as other educational establishments located in the center of the city) and the Teachers' district, which was named for the District Teachers' Training School, whose students were extremely active. The 2nd Gymnasium was represented on the committee by Tamarkin and Friedmann (who were called members of the Central Committee). Both of them, in keeping with the traditions, or perhaps external attributes, which, as we see, the young people quickly picked up from their older comrades, had their own party names: Tamarkin had the code name "Pepov," and Friedmann "Lilovy" (it is now probably impossible to identify their origin!). The book gives pictures of "the most active members of the St. Petersburg social democratic students' organization," the first of whom is A. A. Friedmann (unfortunately, the quality of the picture does not allow us to reproduce it here). It is noted that the Central Committee got together every week on Sunday nights in the apartment of one of its members. We further learn that the organization's Central Committee was directly linked with the St. Petersburg committee of the RSDWP (the Russian Social Democratic Workers' Party) and that it had three hectographs which printed hundreds of copies of leaflets. This is what gives "flesh and blood" to the reminiscences of V. I. Smirnov, given to one of the authors of this book, saying that on the way to the sessions of the mathematical circle Friedmann often called into an apartment house where people working at the Imperial Court lived, to "drop" stacks of leaflets in one of the apartments. Friedmann rightly believed that they would be safe in there. This house is undoubtedly 32 Palace Embankment, where A. I. Friedmann, a doctor's assistant in the Court's Medical Unit, was living. Friedmann's involvement in the work of the leading body in the Social-Democratic Organization of secondary school students (which had its editorial board publishing a handwritten *Voice of Secondary Educational Establishments* and was intending to issue the *Newsletter* of the Students' Organization) allowed us to suggest that the resolutions of the meetings of senior school students of the 2nd Gymnasium were drawn up with most active participation of its two students – Friedmann and Tamarkin.

After graduation from school, Friedmann's and then Tamarkin's political activity declined, as they increasingly concentrated on their studies, as well as on their work in the students' editorial committee, which will be dealt with in the next chapter. But the zeal of the years of the First Russian Revolution was not wasted, and manifested itself in subsequent years in Friedmann's active participation in the organization of Russian military aviation during World War I, in his activities at Perm University, and in his titanic effort aimed at organizing Soviet science in the last five Petrograd years of his life.

3
University years, 1906–14

Undergraduate studies

Friedmann was lucky: in 1906, the year he entered the University, Professor Vladimir Andreyevich Steklov (1864–1926) was transferred there from Kharkov. He was a Corresponding Member and later a Full Member of the Academy of Sciences, and in the Soviet period its Vice-President: one of the organizers of science in the USSR, a brilliant mathematician who continued the best traditions of the St. Petersburg mathematical school, glorified by the names of P. L. Chebyshev, A. A. Markov, A. N. Korkin, A. M. Lyapunov and many others.

Vladimir Steklov played an extremely important role in Alexander Friedmann's life. He was not only an outstanding mathematician, but also had an unusually bright personality. Nature had endowed him with excellent genetics. His uncle on his mother's side was Nikolai Alexandrovich Dobrolyubov, the famous Russian literary critic. His father, Andrei Ivanovich Steklov, a highly educated person, taught history and Hebrew at the Nizhni Novgorod Theological Seminary. Vladimir Steklov had unquestionable literary and musical talents – he could have been a successful opera singer. Steklov has bequeathed to us not only classical works in mathematics and mathematical physics, but also works of some literary merit. These include his book *To America and Back* (Leningrad, 1925), based on his impressions of his trip to the United States, as well as his books about Lomonosov and Galileo. Yet his vast literary legacy is still waiting to be published and commented upon. For almost two decades Vladimir Steklov kept a diary, and the archives of the USSR Academy of Sciences hold colorful notebooks filled with his daily jottings. To read them is to see a whole epoch through the eyes of a contemporary who was also an attentive observer, as well as to see the life of St. Petersburg science, the Academy and the University through the eyes of an industrious and productive scientist.

These diaries served as an invaluable source of evidence about our hero, whose name is often encountered in Vladimir Steklov's diary from 1908 onwards.

Friedmann's autobiography had only four lines dealing with his university years: in 1906 he entered the University, in 1910 he graduated from it, then wrote, jointly with Ya. D. Tamarkin, "An investigation of second-degree indeterminate equations," which was awarded a gold medal. This scanty information is, of course, confirmed by documents, which help us feel the spirit of the time. Let us, therefore, quote the "application" which Friedmann submitted on July 26, 1906 to His Excellency the Rector of St. Petersburg University:

> Submitting herewith my documents, i.e. matriculation certificate from the 2nd St. Petersburg Gymnasium (No. 521), certificate given by the school director (No. 580), birth certificate (No. 51), draft registration certificate (No. 45), and notarized copy of a service list of my father, Gubernsky Councillor Friedmann (of June 27, 1906), and attaching herewith the required fee of 25 rubles payable to the University, I have the honor to beg you to admit me to the first year of the Physics and Mathematics Faculty of the Mathematics Division of the University entrusted to you.
>
> A. Friedmann

So in August 1906 Friedmann (and certainly Tamarkin as well) was enrolled in the University. One can see what lectures he attended in his student's record book – the "record of courses attended" which notes, among other things, that he regularly paid the required fee for attending the courses. Against the name of each course there is an invariable mark: "quite satisfactory," the names of the professors and their signatures. The data about the professors who were giving those courses can, if necessary, be taken from the biographical dictionaries mentioned above. Introduction to analysis and differential and integral calculus were given by D. F. Selivanov, higher algebra by Yu. V. Sokhotsky, application of differential calculus to geometry by I. I. Ivanov, analytic geometry and application of integral calculus to geometry by I. L. Ptashitsky, calculus of finite differences and theory of probability by N. M. Gunter, higher and descriptive geometry by S. Ye. Savich, the mechanical section of physics, theory of electricity, optics and acoustics by N. A. Bulgakov, physics of partial forces and thermal physics, higher optics and thermodynamics by O. D. Khvolson, and virtually all the courses in mechanics by D. K. Bobylev. Against some courses that Friedmann attended in his final years at college his record book has the note "exempted" – probably exempted from tests or examinations because Professors Bobylev, Steklov and Sokhotsky were sure of the results without examinations.

Friedmann in the early 1910s.

Steklov's diaries

Information about Friedmann the student, and documentary evidence of his activities, can be found in V. A. Steklov's diaries. Reading them gives one a vivid impression of their author, and one cannot but feel affection for him. Steklov would always begin his notes with the weather as of ten o'clock in the morning: the temperature and atmospheric pressure, whether the sky was clear, whether it was raining or snowing, etc. Apparently, before going to bed he would write down his impressions of the day: household events; a review of his vast correspondence (to whom he wrote and what he received!); what he had done during the day – whether at his desk in his apartment (first in Zverinsky Street, and later, in 1919, in Vasilievsky Island) or at the University, or at the Academy of Sciences; whom he visited (most often, an established group of St. Petersburg professors – A. M. Lyapunov, A. N. Krylov, L. A. Chugaev, A. A. Markov – and their wives); and who visited him and his wife Olga Nikolaevna Steklova. There were unplanned visits during the day, too. Steklov received, apparently having to interrupt his work, anybody who came to visit him. Some displeasure or grumbling might be reflected in the wording of the note: not "came" but "turned up." The last entries of

Steklov's daily (or rather nightly) notes: 3.30 (in the morning), the temperature is this, the atmospheric pressure is that, the weather is such and such.

One can use Steklov's diaries to study a whole epoch – and what an epoch it was! War, revolution, post-revolutionary years. The description of world events is next to everyday issues, indications of food prices, etc.

However, among all these treasures we were mainly interested in what concerned Friedmann and his friends. Chronologically, the first mention of Alexander Friedmann was found in the entry for January 13, 1908. At that time Tamarkin and Friedmann – third-year students – were attending Steklov's course in the integration of partial differential equations.

Steklov records his impressions in a rapid handwriting, abbreviating some words and often not leaving any space between others (we expand the abbreviated words – they are easy to decipher). Thus:

January 13, 1908. At 4 o'clock Tamarkin and Friedmann (undergraduate students) turned up and brought the continuation of the lectures in integral calculus they had written. They took the ones I had corrected (i.e. looked through. No possibility of correcting them properly!) They said they would come to my lecture on the 16th. They asked me if it was possible to legalize the mathematical society without a supervisor. I told them to make some suggestions. Let us see!

Thus, the two friends were taking down the lectures delivered by Steklov, and after he had looked through their notes they would have them printed lithographically (this is known from other sources). This was a matter of concern to the editorial commission of the Physics and Mathematics Faculty, which will be dealt with later. The faculty had a society of students majoring in mathematics, which was quite independent, with no teachers allowed (a similar situation existed in the society of physics students). A. F. Gavrilov recalls that all the faculties had such students' societies, to which frequent references exist in the published minutes of the sessions of the Council of St. Petersburg University for 1906, and several later years. There were a physics circle; an aeronautics circle; circles for political economy, epistemology, and the philosophy of the state; a circle for literature and art; studies of the foundations of Roman law; lovers of nature and hunting – over twenty different circles with a variety of programs. All of them were united into the University Inter-Society Organization, whose chairman was V. L. Komarov, the future president of the USSR Academy of Sciences, at that time an associate professor at the University. This organization financed the publication of the "Proceedings" of some of these societies in 1909 and later years. Gavrilov writes that Friedmann was an active participant in these societies: "I remember his talk at one of these circles[1] 'On the canals of Mars', the news of whose discovery had been announced not long before. Amid the stormy applause of the young audience Alexander Alexandrovich ended his paper with these words: 'The canals appeared almost suddenly; at any rate, they have been built within a very short time. Could this be evidence that socialism has already been built on Mars?'"

It is a telling episode, characterizing the mood of the students at that time and particularly Friedmann himself (let us recall that in the recent past he had been a member of the Central Committee of the Social-Democratic Organization of St. Petersburg students!).

The students failed to get the mathematics society legalized without a supervisor: Professor D. F. Selivanov was officially appointed, but he never appeared at the meetings, so the students got what they wanted *de facto*, if not *de jure*.

Let's continue with Steklov's notes.

January 25. Began lecturing at the University [after the vacation – V. F.]. Not many students. [I. I.] Ivanov not lecturing at all.

February 20. Brought my collected works to the University and gave them to Tamarkin for the students' mathematical society. Three memoirs are missing.

April 24. Got an invitation from the students' mathematical society to its meeting at 8 o'clock on Thursday. Some student called Doinikova is going to say something about some theorem of Appell's. No time, I'll not go.

We will tell you about Valentina Vitalievna Doinikova later. Paul Emile Appell was a French mathematician and student of mechanics; one of his articles – "On dissipative friction forces" – was discussed by Doinikova together with Friedmann and his group. Doinikova spotted a significant mistake in Appell's article. Friedmann was upset that he had missed the mistake, and remarked half-jokingly: "God teaches infants." Doinikova told me about this episode over 60 years after it had taken place. A surprising memory! Confirmation of oral evidence through documents is always a thrill for a historian! Back to Steklov's diary.

September 10, 1908. Began lecturing at the University on the integration of partial differential equations. There were quite a lot of people. Before the lecture I said a few words about A. N. Korkin[2] and suggested that the students pay tribute to his memory by standing up, which of course they did. Friedmann said that classes could hardly continue and a passive strike was likely to develop by itself.

---

[1] The physics circle.
[2] Professor Alexander Nikolaevich Korkin had died on September 8.

Well, this is not certain. They may decide to delay examinations [note that in those days examinations could be taken throughout the whole year and not during examination sessions as is the practice today – V. F.].

September 13. Lectured on the integration of partial differential equations to third-year students. Not many students, but still some. Meeting today, about a thousand students signed. They want a strike.

September 27. Tamarkin and Friedmann came to see me. Stayed till 11. Brought lectures on integral calculus and equations. They're just boys, real children! They're in raptures over their strike.

The text of A. F. Gavrilov's memoirs seems to be written as a commentary to these notes: "The University was in turmoil, the students spent half their time at meetings, students' strikes accompanied the people's fight against the government. Alexander Friedmann, particularly in his early years at the University, was involved in the students' political movement."

And again Steklov's notes.

October 21. Tamarkin and Friedmann came to see me this evening. They are going to organize a mathematics reading room. Asked me to be their supervisor. Declined, but they deserve help.

November 22. Tamarkin and Friedmann came to see me this evening . . . Kept asking me about their delvings into the theory of orthogonal functions. They are having an article published in Crelle's journal [Steklov means the article "Some formulas pertaining to the theory of the function [x] and Bernoulli numbers" published in the German *Journal of Pure and Applied Mathematics* in 1909 – V. F.]. Sharp fellows! They left at half past twelve, after supper.

April 18, 1909. The students Tamarkin, Friedmann, Petelin came to see me this evening . . . I proposed to Tamarkin that he think about the asymptotic solution of differential equations (i.e. stability, in the sense of Poincaré and Lyapunov) or the problem of equilibrium of a rectangular plate. To Friedmann I suggested he find orthogonal substitutions, when fundamental functions are products of two (see my dissertation).[3] I suggested Petelin read what Jacobi had to say about the principle of the last multiplier. I'll think it over again and will probably find some other topics too.

April 25. Finished my lectures on integration of equations. For some reason, Messrs. Students greeted the end of my lecture with applause (were they happy I was through?).

September 12. This evening Tamarkin, Friedmann and Petelin came to see me. They had worked on the assigned topics. Seem to have done something. Promised to submit their essays in a month. Tamarkin seems to be doing better than the others.

[3] This work by Friedmann will be dealt with under "Postgraduate studies" later in this chapter.

Friedmann and Ehrenfest

Along with V. A. Steklov, another strong influence on Friedmann was Pavel Sigizmundovich Ehrenfest, whose name is very much part of the development of theoretical physics in Russia in the pre-revolutionary and early post-revolutionary years. At the end of 1907 or the very beginning 1908, the many societies active at the University and other higher and secondary educational establishments in St. Petersburg were supplemented by a "kruschok" whose work was supervised by P. S. Ehrenfest. We deliberately wrote the Russian word "kruzhok" (a hobby group, a society) in German – the way Ehrenfest wrote it, for Russian was not his native language. Paul Ehrenfest was born in 1880 in Vienna and lived there during the first half of his life. He entered the Faculty of Philosophy at Vienna University when the great Boltzmann was a professor there. Ehrenfest was the latter's student and later a close disciple. Under Boltzmann's guidance he wrote a paper on classical mechanics, for which he was awarded a Ph.D. Ehrenfest decided to continue his education in Germany and to attend the lectures of the most outstanding mathematicians of the century – Felix Klein and David Hilbert. In Göttingen in 1903, at one of the lectures, he met Tatiana (Tanya) Afanasieva, who had graduated from the famous Bestuzhev Women's Courses in St. Petersburg and had come to Göttingen with the same aim as Ehrenfest. In 1904 the young couple married, and in the autumn of 1906 they came to live in St. Petersburg, Tanya's native city. There, within a short time, Ehrenfest became intimate with the leading young physicists, particularly A. F. Ioffe and D. S. Rozhdestvensky. Towards the end of 1907, Ehrenfest's "kruschok," or modern physics seminar, was already meeting in Vasilievsky Island, where the Ehrenfests had settled (and, later, in Lopukhinsky Street, in the Petersburgskaya Storona). Besides the above-mentioned young Russian colleagues of Ehrenfest (and his wife!), among its active participants were K. K. Baumgardt, V. R. Bursian, G. G. Weichardt, L. D. Isakov, Yu. A. Krutkov, G. Ye. Perlitz and V. M. Chulanovsky, teachers at St. Petersburg University or graduates who had stayed on at the University to prepare for the teaching profession; V. F. Mitkevich, an electrical engineer from the Polytechnical Institute; V. V. Doinikova, M. A. Levitskaya and A. B. Feringer, physics students from the Bestuzhev Women's School; and finally – Ya. D. Tamarkin and A. A. Friedmann, mathematicians, who are of special interest to us. The names of Friedmann and Tamarkin are found in the correspondence between Ehrenfest and Ioffe; V. V. Doinikova told the author in detail about

Yuri A. Krutkov.

Friedmann and his relations with Ehrenfest. It is important that Ehrenfest's archive kept in Leiden (Netherlands) contains Friedmann's letters to Ehrenfest, considerably supplementing the information about Alexander Friedmann that was at our disposal.[4]

At its weekly meetings, Ehrenfest's circle discussed primarily the most recent topics in quantum theory, the theory of relativity and statistical mechanics. The papers were of two kinds, original papers and abstracts, but even the abstracts were highly creative. Ehrenfest highly valued, of course, the ability of his colleagues to capture the main points in the work reviewed and to present them clearly to others. But he encouraged the participants in the seminar not to worry if they failed to understand certain (usually very complicated) questions, but to emphasize these problems and to clarify them through free discussion at the circle's sessions. Doinikova recalled that Ehrenfest would invariably and immediately ask everybody who came to the seminar in Lopukhinsky Street: "Well, what questions have you got?" Once, this traditional query produced the reply: "Why, Pavel Sigizmundovich, can't we come and visit you unless we have questions?" The surprised Ehrenfest objected: "What do you mean, unless you have questions? If you study physics you can't *not* have questions!"

Unfortunately, we do not know the exact titles of the papers presented by Friedmann in Ehrenfest's circle, but we know that he became very

---

[4] These letters were made available to us through the courtesy of Dr. A. I. Engberts (Netherlands) and Dr. Josephson (USA).

intimate with the Ehrenfests and frequently visited them – most often with Tamarkin. Indeed, in the first two decades of this century Friedmann and Tamarkin were regarded almost as a single person. Thus, for the Ehrenfests' little daughter Tanya (who, in keeping with the family's mathematical traditions, was called Tanya-dashed, to distinguish her from her mother, for whom she was named) the two young men were "the thin one and the fat one," the thin one being Alexander and the fat one Yakov, who even in his youth had a tendency to put on weight. The same Tanya-dashed, as Doinikova recalled, sometimes called the two friends "little spider" and "little elephant." Friedmann had long thin fingers, which is not surprising for the son of musicians – hence apparently his nickname.

Let us now quote some parts of Friedmann's letters to Ehrenfest dating from 1908–10 and briefly comment on them.

Tim, May 14, 1908.

Dear Pavel Sigizmundovich,

I am writing this message in block letters for your convenience, and since I'm not skilled in this art, something unreadable seems to be coming out. Generally speaking, I like it very much here, in the town of Tim. But there is still this nagging boredom, from which only my cargo of books can save me. But the cargo won't arrive; there is no certificate for it, and Tamarkin, on top of that, has vanished into thin air. Should you meet him, please ask him if he has sent me my holiday permit; unless he sends me this document, I shall have to go to prison here and then back to St. Petersburg under military escort. Write and tell me your news. Apologies for this wretched letter, I've not yet recovered from the nervous strain.

I remain,

Respectfully yours,

A. Friedmann.

P.S. Send me a short reply. Everybody has forgotten me and I'm depressed; write either in German or in Russian as you wish. A. Friedmann.

This letter is interesting right away for the place where it was posted. Its name will not mean anything to most readers. In the recent (Russian) edition of *The Dictionary of Geography* (1986) the town of Tim is not mentioned, though can be found on the little map of the Kursk Region placed in the dictionary. The town is 60 km east of the city of Kursk, and even today it is not on the railroad. The very detailed encyclopedia of Brockhaus and Efron (vol. 65, published in 1901, just before the events described) does not fail to mention Tim, though it says that it "has no commercial or industrial significance." In 1907, its population was 7,380 (3,512 men and 3,868 women). So, in the spring of 1908, its population increased by one gentleman! But, what is more important to us, Valentina Doinikova (and her parents) were living in the town in those years – it is she that Alexander Friedmann was visiting. "The best nightingales are in the Kursk Region and the best of them are in Tim!" she recalled in 1969. When she was speaking about Ehrenfest (nothing gets lost – manuscripts do not burn! – there is Doinikova's letter to the Ehrenfests in which she invites them to come and visit her in Tim and gives detailed instructions: at what station they were to get off, how long it will take them to get to Tim), she mentioned Friedmann's name, and I asked her to try and recall what she knew about him. Doinikova began to talk and suddenly stopped, and her wrinkled face flushed. "I must tell you something before I go on with my story. Alexander Friedmann was my fiancé. Then our engagement was broken off, but we remained on good terms." Valentina Doinikova continued her story, and she spoke about Friedmann with great warmth. Over 20 years have passed since that conversation. I have learned much about Friedmann since that time and can only be surprised by the accuracy of Doinikova's account of Friedmann's and her years at the university, about which I learned something later from papers and letters dating to those distant times. "Friedmann, Tamarkin and Smirnov often came together, and they were called 'the boys from the 2nd Gymnasium.' They were always smart and neatly dressed, and always called each other – in public – by their first name and patronymic. Alexander Friedmann usually wore a bowler hat and always carried an umbrella. And he didn't part with his umbrella when he was in the Kursk Region, in Tim, where everything was lit with the sun in the summer. There was a young man named Andrei Borisyak, nice and cheerful – always the life and soul of the party. Incidentally, he discovered a supernova one day, for which Nicholas II granted him a telescope. Boris wrote verses and drew well. I can still see his friendly caricature: Friedmann walking with a bowler hat on and an umbrella. Friedmann was full of contradictions: fanaticism, for instance, combined with unusual pragmatism. He was bold, but I remember us riding on horses in Tim – Friedmann was afraid of them. He was a typical townsman, who found horses more than he could cope with."

I wrote down a few words that she said about Alexander Friedmann's literary tastes (I wish now I had put down more, but my task was to find out as much as possible about Ehrenfest, and as to Friedmann, there seemed to be enough time . . . Alas!). His favorite writers were Dostoyevsky, E. T. A. Hoffmann, and Poe; he read a lot of Merezhkovsky and Fyodor Sologub. Indeed, Friedmann, as it turns out, was fond of

V. V. Doinikova, A. F. Ioffe and "Tanya-dashed" Ehrenfest in
Kannuka, 1909.

"the occult" in general; he claimed he could soothe pain, and once
managed to do it. Passion for science – that was natural, but what
Doinikova said – that Friedmann was a very lively and witty partner in
conversation, in fact liked to play the fool, and made it seem charming
and funny – was somewhat unexpected.

As Valentina Doinikova recalls, in his university years Friedmann was
particularly fond of Borel's theory of sets, Jordan's *Cours d'Analyse* (in
three volumes) and Goursat. He revered V. A. Steklov, spoke about A. A.
Markov with some comic awe, and highly esteemed Ya. V. Uspensky,
whom my notebook recording this conversation with Doinikova calls
"the closest of the idols." What she meant by this epithet, one cannot even
guess, let alone establish with certainty! And one more note: "According
to Friedmann, in choosing one's specialty one should be guided not so
much by the substance of the discipline [apparently its facts and results –
V. F.], but by its methods. Skilled craftsmen should go into physics,
people inclined to abstract thinking should go into mathematics. One
should not be afraid of calculations." And he was not afraid of calculations. "He could sit day and night covering hundreds of sheets of paper

with figures," Doinikova said. "In this he was like my husband, his colleague and friend A. S. Bezikovich. Ehrenfest said of Bezikovich: 'That man is a machine, there's no problem he couldn't solve. A bull of a man!'"

"We remained friends with Friedmann after the University too when we were in Perm, and then again in Petrograd. He presented me with a book in which he wrote: 'To V. V. Doinikova from a friend in mathematics'."

We skip over here some of Friedmann's other letters to Ehrenfest. They contain a discussion of some mathematical questions which interested Friedmann, particularly those concerning the theory of functional equations,[5] as well as critical comments on Khvolson's general course of physics. They also contain philosophical speculations (sometimes rather naive) about the role of axiomatics in mathematics, physical science and even literature. Reading Friedmann's letters and documents, one thinks about him primarily as a founder of relativistic cosmology, forgetting that in 1908, during his rest (and work!) in the town of Tim, he was quite a young man!

St. Petersburg, August 18, 1908.

Dear Pavel Sigizmundovich [Ehrenfest],

This letter will be brought to you by Tamarkin. When he came I learned from him that thanks to the publishing committee I turned out to be a real swine and dishonest person, not having given you the money. For God's sake, forgive me for this oversight. I would, of course, have apologized to you earlier, if only I had known that the debt had not been paid to you . . . I would come to see you with pleasure,[6] but I have no spare money at all and although, nominally, I will have a large sum of money this week, actually, it will not belong to me, and the trip to you will cost about ten rubles.

My present state of mind is melancholy at least squared.

I apologize again for the unintentional meanness I have committed. Give my regards to Tatiana Alekseyevna, your aunt, and Tatiana Pavlovna.

I remain,

Respectfully yours,

A. Friedmann.

My new address: 17 Sergievsky St., Apt. 21.

The publishing committee (PC), mentioned in this letter, deserves a little more detailed treatment. Friedmann's active involvement in its work is mentioned by A. F. Gavrilov. The committee's activity was regulated

[5] The simplest case of such an equation has the form $f(x) + f(y) = f(x + y)$; it has the solution $f(x) = Ax$, where A is a constant.

[6] The Ehrenfests were living in the little village of Kannuka located on the seashore between Ust-Narva and Sillamaa.

by its constitution, which stated that "the object of the committee is to publish lectures, with the consent of professors, on those scientific disciplines which are taught in the mathematics section of the Faculty of Physics and Mathematics[7]." The choice of lectures, negotiations with professors, choosing compilers of lecture courses, and business contacts with copyists and the publishers were taken care of by the bureau of the PC, which consisted of five persons, including the chairman, secretary and treasurer. The financial side of this work was of considerable importance and was specified in detail in the PC's constitution. Remuneration for the preparation of lecture courses for publishing could undoubtedly be of help to students, and was nine rubles per signature of lithographed text for the first printing; five and three rubles for the second and third respectively; and beginning with the fourth printing the income went to the PC's treasury. Another two rubles were given for copying a signature. The copyists' handwriting was excellent! The editors were to be paid – unless their work was done by the authors. The activity of the PC, on which its bureau reported to the students twice a year, was audited by the auditing commission; everything was like in the adult world! It was, in fact, a quite mature organization to which the Physical Institute of the University assigned a special room. The courses were printed in Bogdanov's printing house; the PC bureau would send obligatory copies to the public library (three copies) and the university library (three copies). In addition, three copies were given to the professor who was giving the course, and one copy to the compiler.

The courses published by the PC have been preserved, but unfortunately without the names of the compilers, the price, or the circulation. According to Gavrilov's memoirs, Friedmann compiled the courses based on "Differential Calculus" by Professor D. F. Selivanov (at least four printings), "Integration of Functions" by the same author (three printings), "Higher Algebra" by Yu. V. Sokhotsky (four printings), and "Theory of Numbers" by I. I. Ivanov (compiled jointly with Tamarkin). It cannot be that he did not take part in publishing Steklov's courses, particularly his "Partial Differential Equations," reprinted by the PC many times. As to Ehrenfest, he may have been involved in editing some courses, and there must have been some misunderstanding over his pay for the work, which was taken by Friedmann so close to heart.

Dear Pavel Sigizmundovich,

Excuse me for troubling you with my small request. You may know what

---

[7] *The Minutes of the Sessions of the Council of St. Petersburg University in 1907*, St. Petersburg, 1908, pp. 345ff.

mathematical journals are published in Germany, Austria, and France for secondary school students in elementary mathematics. If you do, I would be grateful if you could let me know. I would like to contribute an article about journals for secondary educational establishments to *The Russian School*. That's why I need this information. My financial situation is fairly bad, and I am now earning a little money from *The Russian School*.

I remain,

Respectfully yours,

A. Friedmann.

Give my regards to everybody. I will be very grateful for a reply. My address: 17 Sergievsky St., Apt. 21.

Unfortunately, this letter of Friedmann's bears no date. It probably dates from 1909 or 1910. It is clear that Alexander Friedmann was in strained financial circumstances. It seems that his father had died (the date of his death has not been established; from conversations with V. V. Doinikova it is only known that he had died before Friedmann's grandfather, i.e. before 1910). That is why Friedmann moved to his grandfather's place in Sergievsky Street (now Tchaikovsky Street). His grandfather had been retired for a year already, so it became awkward for his grandson to abuse or, perhaps, use his support at all. In search of an income he applied to *The Russian School*, a weekly published by the famous St. Petersburg pedagogue Ya. Ya. Gurevich. Having looked through the journal, we found in the 1908–10 issues a few reviews signed by "A. F—n." However, one cannot be sure that our "Friedmann" is behind all of them. In fact, in Gurevich's journal there was a contributor, A. Fomin by name, who sometimes signed with his full name, but sometimes also with initials. Judging by the content of some of the books reviewed in the journal, which could not possibly interest Friedmann, those reviews were written by Fomin. However, Friedmann's authorship can only be conjectured, because there are some reviews which could have been, in principle, written by him, yet are signed by Fomin. But here is a review of S. Melgunov's book *Students' Organizations of the 1880s–1890s at Moscow University (Based on Archives)*, Moscow, 1908, 50 kopecks. This is almost certainly Friedmann, with his interest in student (and generally youth) movements, and his involvement in university life. "The book," says the review, "will be read with interest not only by people who are specially interested in the history of our universities, but by everyone who is interested in the history of our society, of our social movement. The Russian university, due to historical conditions, has always been the social barometer."

Another review is that of a French book (not translated into Russian) by S. Santerre *Psychology of Number and Elementary Operations in Arithmetic*, Paris, 1906. This book is devoted to the philosophy of mathematics, to its new trends; the thrust of the book (judging by the review) is the priority of mathematics in the alliance of sciences, and the author is striving to unite it with psychology. Friedmann might ordinarily have reviewed two other books: one by E. C. Brewer – *A Guide to the Scientific Knowledge of Things Familiar*, which had had 46 (!) editions in England, and one by H. Weimer – *A History of Pedagogics*. But Friedmann's reviews were no longer being published in the journal, mainly, it seems, because he had talented rivals in the long-standing contributors N. Tomilin (who reviewed books on physics) and F. I. Pavlov (mathematics).

However, there is reason to believe that Ehrenfest did not ignore Friedmann's request. Since 1908, Ehrenfest had been very actively involved with the *Journal of the Russian Physico-Chemical Society* (the physics section) and *Problems in Physics*, the journal's supplement. In these supplements we found several contributions by Friedmann published in 1910, including short reviews of V. F. Kagan's article "Mathematics" from the book *History of Russia in the First Half of the 19th Century*, and of *Proceedings of Students' Scientific Societies in the Physics and Mathematics Faculty of St. Petersburg University*. There is also a more detailed review of Lucas de Pesloüan's *Abel: his Life and Work* published in Paris. Like the author, Friedmann writes with great sympathy about the hard fate of this outstanding Norwegian mathematician, who was not understood even by his great contemporaries Cauchy, Legendre and Gauss ("It is terrifying to read about the loneliness that oppressed Abel," wrote Friedmann). The review contains a statement by Friedmann about books of this kind, which is important for his biographers: "In the life of a scientist," wrote the 23-year-old Alexander Friedmann, "besides his activity as a thinker, one should always distinguish two significant aspects: material conditions and private life." Concerning the former part of Abel's life, Friedmann notes that "the great scientist had to be content with 40 kopecks a day – a sum which at that time was entirely insufficient." And Abel was already a university professor! The article about Abel's biography is written by a mature person, a perceptive and discriminating reader.

Again in 1910, *Problems in Physics* published an article by S. A. Dianin and A. A. Friedmann "On the return of Halley's comet" – in that year the comet was expected to pass through its perihelion, and the coming astronomical event stirred many minds (the spectrum of the comet was to be studied for the first time).[8]

And one more observation concerning the *Journal of the Russian Physico-Chemical Society* for 1910–11: the title pages of volumes 42 and 43 bear a laconic inscription: "The proofs read by A. A. Friedmann" (in 1911 he shared this responsible task with L. D. Isakov, a participant in Ehrenfest's seminar, and F. F. Sokolov) – further evidence of Friedmann's reduced circumstances.

Postgraduate studies

The university years, filled with attending lecture courses, preparation for examinations, involvement in student research groups, and joint studies with Tamarkin, quickly flew by, and in 1910 the university course was completed. Friedmann had to think about what to do next. The best option for him (and his mathematician friends) was to stay on at the University to prepare for a teaching position – the equivalent of today's graduate studies. Luckily for Friedmann and his classmates, this idea was given full support by V. A. Steklov. Correspondence on this issue with the superintendent of the St. Petersburg educational district was started in the summer of 1910. Those who were retained for the training program were divided into two classes – "salaried" and "unsalaried." In that year, among the graduates were Tamarkin, Friedmann and another very able student, V. V. Bulygin. So, recommendations were sent for all three, and V. A. Steklov also wrote references for all three (he was sometimes joined by Professor D. K. Bobylev); therefore, we had to look for these references in the files of Bulygin, Tamarkin and Friedmann. It often turned out that the beginning of the reference was in the file of one of them, and the continuation in the other files. References and recommendations had the same format. Here is an example of such a recommendation addressed to the superintendent, which concerns the hero of the book:

I have the honor to apply to Your Excellency in compliance with Art. 427, Sec. IV, Item 4 of The University Charter of 1884 concerning the retaining at St. Petersburg University, in the department of mathematics, of Alexander Alexan-

---

[8] It is to be noted that for some reason all the above-mentioned contributions by Friedmann, including the article on Halley's comet, are signed with initials (as are the reviews in *The Russian School*); in this case the initials "L. T." were chosen, which do not seem to admit of logical decoding. At the same time, in the contents of volume 42 of the journal the last names and the initials of the authors, A. A. Friedmann and S. A. Dianin, are given in full. Finally, it should be noted that all these publications of Friedmann are missing from the published bibliography of his works.

drovich Friedmann, chosen by the Physics and Mathematics Faculty to prepare
for professorial work for two years: (1) from September 1, 1910 till January 1,
1911, and concerning the grant to him of a stipend of 400 rubles from September 1, 1910 till January 1, 1910. Enclosed are eight documents, a first degree diploma . . .

Let us now quote the official reference given by Steklov and Bobylev
(Friedmann's file contains a copy), which is written in the name of
Steklov:

Until recently Mr. Friedmann has been working together with Mr. Tamarkin,
and all that has been said about the latter's school years and first three years at the
University is fully applicable to the former, i.e. to Mr. Friedmann. The above-
mentioned papers [there is a gap in the text of the copy; the original apparently
had the titles of two published papers by Tamarkin and Friedmann on Bernoulli
numbers – V. F.] were written jointly by Mr. Friedmann and Mr. Tamarkin, as
well as an essay which was awarded a gold medal. The two have equal abilities and
already give the impression of young scientists, rather than recent graduates.

In his last year at the University he was working on an essay on the subject I
assigned: "Find all orthogonal substitutions such that the Laplace equation,
transformed for the new variables, admits particular solutions in the form of a
product of two functions, one of which depends only on one, and the other on the
other two variables" [see Steklov's note in his diary of April 18, 1900 – V. F.]. I
touched on this problem in my doctoral thesis, but did not treat it in detail, having
different objectives to pursue. Though aware of the complexity of the question,
which is a generalization of a similar problem [there is a gap in the text – V. F.],
nevertheless, I suggested that Mr. Friedmann should try to solve this problem, in
view of his outstanding working capacity and knowledge compared with other
persons of his age.

In January of this year, Mr. Friedmann submitted to me an extensive study of
about 130 pages, in which he gave a quite satisfactory solution of the problem.
The paper is a quite independent study and supports the hope that its author will
make with time a talented and independent scientist.

Mr. Friedmann has a good command of French and German and reads English
and Italian papers with ease, which is shown by his own work, and is also known
to me by his correspondence with an Italian scientist [there is gap here – V. F.],
who worked on the same problem as Mr. Friedmann did in the above-mentioned
paper (where Mr. Friedmann points to an inaccuracy in the investigation [there is
a gap here – V. F.)].

The original has been signed by Professors D. K. Bobylev and V. A.
Steklov.

Here is an extract from a similar reference contained in Tamarkin's file.
The reference deals first with his research at school and during his first
years at the University, and continues:

In view of the above I allow myself to ask the Faculty to endorse my recommendation to retain all three above-mentioned persons [i.e. Bulygin, Tamarkin and

Friedmann – V. F.] to prepare for the position of professor of mathematics. I must add that during the twenty years of my teaching career I have never met young people as gifted as these three. They should be retained for science's sake at any cost and should be given the opportunity to dedicate themselves to scientific work without distracting themselves by external pursuits. It is to be noted that one of them, Mr. Friedmann, is without means and when at the University made his living by private tutoring, proofreading and earning a small salary. Messrs. Bulygin and Tamarkin, although having relatives, cannot, of course, continue to be supported by them.

I request the Faculty to apply for a stipend of 1,200 rubles a year for two years from the Ministry's funds for each of them.

Since these stipends, as far as I know, cannot be paid before the end of the current year of 1910, or even the beginning of 1911, and Messrs. Tamarkin and Friedmann (especially) and Bulygin will not be able to live for six or seven months without a permanent income, which will negatively affect their scientific work, I would ask the Faculty to apply for them to be retained at the University as soon as possible with a stipend of a hundred rubles a month to be given to them before January 1, 1910.

Prof. Steklov.

I support the request. Prof. D. Bobylev.[9]

This is a telling document giving an idea not only of the abilities of Bulygin, Tamarkin and Friedmann, but also of Steklov's outstanding personal qualities.

Although the above citations give a quite convincing characterization of Friedmann and his colleagues, we cannot resist quoting several other phrases from a similar reference given by Steklov, through whose hands many dozens of students passed. We read in the file of his pupil M. F. Petelin: "I should note that the class of 1910 is exceptional. In the class of 1911 and among the fourth-year students who are about to graduate there is no one equal in knowledge and abilities to Messrs. Tamarkin, Friedmann, Bulygin, Petelin, Smirnov, Shokhat and others. There was no such case during the fifteen years of teaching at Kharkov University either. This favorable situation should be used for the benefit of the University" (1912). To the young mathematicians mentioned by Steklov one should add the names of A. S. Bezikovich and A. F. Gavrilov – the former was younger than, and the latter was about the same age as, "the magnificent six."[10]

9 A brief explanation: according to the rules which existed at the time, graduate studies began at the beginning of a calendar year and not of an academic year. In view of Friedmann's strained financial situation, Steklov attempted (encountering difficulties, but finally succeeding) to secure a stipend for the first four months of the 1910–11 academic year for Friedmann and his two colleagues. It is to be noted that a professor's salary at the time was about 300 rubles per month.
10 For the sake of completeness we should mention one more of Steklov's pupils, although he was not part of the fraternity of the above-mentioned mathematicians and

N. N. Andreyev, with whom Friedmann collaborated closely in Moscow at the Aviapribor (aviation instruments) plant, contributed to *Priroda* (Nature) magazine in 1967 his reminiscences about his meeting with Alexander Friedmann.[11] He described our hero as he first saw him in late autumn of 1910, i.e. after the University (while Andreyev himself was still about to apply to the University). Friedmann happened then to be sharing an apartment with the artist A. K. Voskresensky (who seems to have rented a room in the apartment in Servievsky Street). Andreyev was coming to see Voskresensky on business. This is what he wrote: "It was morning, the door was opened by a very pale, almost transparent, tall, lean young man, with unusually intense eyes used to spectacles (he had astigmatism), with a towel in his hands. He said in a spiritless sleepy voice that Voskresensky was not in, and, having heard my name, he drew me into his room keeping his hand in mine and immediately began to discuss the university news. It was Alexander Alexandrovich Friedmann. He spoke very warmly and with great respect about the university professors, though not without humorous comments, and as to V. A. Steklov, he spoke about him with particular admiration and with an elation which I could not explain at the time. He talked about the students' everyday conditions, about 'fraternities,' the ads of which were hanging all along the university corridor, remarking that one should and must attend the lectures of only those professors who can emotionally involve students with their subjects. He inquired about my financial situation, since students at the time paid 50 rubles a year for tuition and with some very rare exceptions there were no stipends. He wrote down my address. He said that he had stayed up too late because something wasn't working out, so he got up late, not having had a good rest. On his desk, as it should be with scientists, there were open books and sheets of paper covered with notes."

The autumn of 1910 was financially the most difficult time for Friedmann. Steklov had not yet succeeded in getting a "graduate student" stipend for him for the last four months of 1910. We have already seen that, to make a living, Friedmann did various kinds of literary work. Another example of such work was found in an interesting publication

collaborated with Steklov somewhat later, in 1917–18. This was Sergei Alexandrovich Dianin who has already been mentioned in this book and who took part together with Friedmann in school political activity. In the 1930s he worked in Leningrad in the Combine of Physico-Technical Institutes – in the theoretical department of the Musical Acoustics Institute.

[11] Their author was not only a namesake of Academician Nikolai Nikolaevich Andreyev (who also seems to have known Friedmann), but had the same initials.

undertaken on the initiative of Professor V. K. Lebedinsky, which has been forgotten by historians of science. During the period in question six issues were published which were devoted to the physics of electro-magnetic oscillations and waves. The issues contained translations of classic works by H. Hertz, Lord Kelvin, W. Feddersen, and others. They also had reviews written by Russian authors. The second issue of this publication[12] contained an article by A. A. Friedmann "On the integra-tion of linear equations of the second degree" (the case of an equation describing processes occurring in the oscillation contours). The article was methodical in character and extremely clear and precise. It thoroughly discussed some propositions which are intuitively felt to be unquestionable, and which we are so much used to that we do not think of the necessity of their strict substantiation. The article is also interesting in that it can be considered as a prologue to the original investigation undertaken by A. A. Friedmann and V. I. Smirnov two years later and published in 1913 in the *Journal of the Russian Physico-Chemical Society (Physics Section)*.

The "mathematical academy"

According to the regulations, the graduates who were retained "to prepare for professorial work" had to submit annual reports on their work. This consisted of studying the literature – as part of the preparation for Master's examinations – and independent investigations. Starting with the very first year (1910–11) Friedmann's report invariably featured his "involvement in the circle formed to study some basic courses in mathematical analysis and mechanics." The circle was formed in 1911. Friedmann names its participants: A. F. Gavrilov, who was at the time in his final year at the University, M. F. Petelin, V. I. Smirnov, Ya. D. Tamarkin, Ya. A. Shokhat – all of them upon graduation were retained at the University. They were later joined by A. S. Bezikovich and V. V. Bulygin. The participants chose as the subject of their studies in the reported year the works of Lyapunov on stability, and of Poincaré on celestial mechanics and the three-body problem. In the next report, for 1911–12, there is more detailed information about the circle: it is men-tioned that it was set up on the initiative of Gavrilov, Tamarkin and Friedmann himself. There were about 30 sessions, i.e. they were held weekly (except in the vacations). Gavrilov gave a course of lectures on the

---

[12] See *Electrical Oscillations* (in Russian), Issue 2. Ed. by V. K. Lebedinsky (St. Petersburg, 1910, pp. 109–122).

theory of surfaces (basing it mainly on the work of Luigi Bianchi, an Italian mathematician); Smirnov gave lectures on the theory of algebraic equations involving mainly the work of E. Goursat and P. Appell.[13] Tamarkin told his colleagues about potential theory, Friedmann about the theory of elasticity (mainly according to Clebsch, author of *The Theory of Elasticity of Solid Bodies*), and Shokhat about the theory of spherical functions (based on the work of the German mathematician Heinrich Heine). At the end of the summer, the report says, the circle was joined by M. F. Petelin.

Shokhat formulated the tasks of the circle somewhat differently, emphasizing the discussion of "difficulties that the circle members encountered in their independent studies." In 1911–12, there were 32 sessions held, and Shokhat singles out the course given by Tamarkin on the theory of trigonometric series and Sturm–Liouville theory. In that year Petelin, Smirnov, and Tamarkin were particularly active during the sessions, 30 sessions being devoted to Goursat's books (Petelin spoke 5 times, Smirnov 9, Tamarkin 5 and Friedmann 11 times).

Acquaintance with the brief reports about the young mathematicians' studies in their circle makes it possible to draw an analogy between the circle and Einstein's "Academy Olympia" which had assembled in Bern ten years before. What we saw in St. Petersburg was a genuine "mathematical academy," whose participants not only studied the major fields of mathematics, but also discussed its basic philosophical problems and foundations.

Looking at Friedmann's annual reports for 1911–13, one cannot fail to be strongly impressed by the working capacity of this indefatigable student. Looking at the literature he studied in different areas of mathematics and mechanics, one is highly surprised by the amount of material he assimilated. The lists of courses, monographs, and memoirs are, I would say, simply frightening.[14]

I believe that the phenomenon of the "mathematical academy" which functioned at St. Petersburg University should be the subject of specialized scrutiny. We will confine ourselves to brief biographical notes on the participants of the circle – all of whom closely collaborated with the hero

[13] It should be noted here that the circle members had a high regard for Goursat (after the October 1917 Revolution his course was translated into Russian and published in Russia).

[14] Almost all the courses and books – there were several dozen of them – in mechanics and the various sections of mathematics mentioned in Friedmann's reports are in his scientific library, which has been preserved in the Main Geophysical Observatory named for A. I. Voeikov. The library has over a thousand volumes.

of this book outside the work of the circle, and some of them after university. At the same time, almost nothing is known about many of them. Thus, I remember that in 1987, having started to make inquiries about some of the above-mentioned people, I phoned Olga Alexandrovna Ladyzhenskaya, a pupil of V. I. Smirnov, and a Corresponding Member of the USSR Academy of Sciences. I told her I was interested in Friedmann's school and university years and was about to start asking her questions when Olga Alexandrovna, anticipating my question, said: "I suppose you want to find out something about Petelin?" – "Quite right!" – "Alas! At that time we did not ask Vladimir Ivanovich [Smirnov] about him, and lost track of him."

The "tracks" of all these university colleagues of Friedmann's emerge mainly from the documents kept in Pskovsky Street, in the Leningrad State Historical Archives. Let us give very brief notes on these mathematicians.

Vasily Vasilievich Bulygin

Born into the family of a military serviceman on May 28, 1888. Upon graduation from the 3rd St. Petersburg Gymnasium (with a gold medal) in 1906, entered the mathematical section of the Physics and Mathematics Faculty of the University. Retained at the University in 1910 to prepare for professor's position. In 1915 became an official in the government institution supervising orphanages and charity work ("His Majesty's Own Office Overseeing Empress Maria's Institutions"), working at the same time as a junior assistant in the department of mathematics at the University and in the Institute of Railway Engineering. Together with A. A. Adamov and others (including Smirnov, Tamarkin and Friedmann), he published a collection of problems in higher mathematics, the fifth (last) edition of which came out in 1931. Died in January 1919.

Abram Samoilovich Bezikovich[15]

Born in Berdyansk on January 24, 1891. Graduated from the mathematical section of the Physics and Mathematics Faculty of St. Petersburg University in 1912. His teacher was A. A. Markov, but V. A. Steklov also played an important role in his life. Retained at the university to prepare for professor's work, Bezikovich became a university instructor in 1916, then a professor, and for some time was Rector of Perm University. In 1920 he came back to Petrograd and taught at the University and the Pedagogical Institute. As a Rockefeller Fellow spent two years, from 1924

[15] This is the exact transliteration; when living in the West he spelt his name Besicovich.

to 1926, in Copenhagen with Prof. H. Bohr. Remained abroad, moving to England, where he taught at several universities, longest of all at Cambridge University. In 1934, became a Fellow of the Royal Society. Widely known for his work in the field of almost periodic functions, probability theory, geometry, etc. Died in England in 1970.

Alexander Felixovich Gavrilov

Born on August 7, 1887, in Warsaw, where he left high school with honors and entered the University. In 1907, he transferred to the mathematical section of the Physics and Mathematics Faculty of St. Petersburg University, where in 1912 he was retained to prepare for professor's work. In 1914, joined the Army; from 1915 to 1918 was involved, together with Friedmann, in the development of bomb-targeting and in the drawing up of the appropriate charts, and in the organization of an aviation instrument plant in Moscow. Then taught at Perm University (1918), and from 1918 to 1920 in Nizhni Novgorod, where he became a mathematics professor and a Deputy Rector. On his return to Petrograd, he worked simultaneously at the Polytechnical Institute (1920–24), at the Chemico-Pharmaceutical Institute, and at the Electrotechnical Institute of Communication, as well as at the Naval Academy (1922–24). The author of a number of courses in mathematics, wrote one of the most substantial articles about Friedmann, first published in 1959. After World War II taught at Leningrad institutes of higher learning. Died in 1961.

Mikhail Fyodorovich Petelin

Born on October 14, 1896 into the family of a non-commissioned officer, a trumpeter in the Household Cavalry. Studied at the 2nd St. Petersburg Gymnasium and "displayed exceptional interest in all the subjects of the school curriculum," as his matriculation certificate says. Petelin's school file has also this note: "For excellent achievements and because of his parents' low income was exempted from tuition fee." Left the Gymnasium with a gold medal in 1905, and in the same year entered the mathematical section of the Physics and Mathematics Faculty of St. Petersburg University. Was retained at the University in 1910 to prepare for professor's work. Friedmann's co-author in one paper (1913). At the end of 1915, began to work in the main Physical Observatory (on V. A. Steklov's recommendation).

Petelin compiled a book *Integration of Ordinary Differential Equations* based on his teacher's course of lectures; Steklov remarked that in the twenty years of his activity it was the only course prepared for publication

Mikhail F. Petelin.

by a student which did not require any editing. The last item of information we have about Petelin's career is his work with V. A. Shokhat at the Mining Institute in Ekaterinburg in 1918; in the same year he returned to Petrograd and soon (in 1921) died. The exact date of his death could not be established.

Vladimir Ivanovich Smirnov

Born on June 10 (New Style), 1887, in St. Petersburg. Studied at the 2nd Gymnasium (in the same class as Petelin). After graduation (with a gold medal) entered the mathematical section of the Physics and Mathematics Faculty of St. Petersburg University. Graduated in 1910, and was retained at the University to prepare for professor's work. Friedmann's co-author in one scientific article (1913), and author of an interesting article about Friedmann. Smirnov wrote, jointly with Tamarkin, the first volume (the first and second editions) of *A Course in Higher Mathematics*, and with Bulygin, Tamarkin and Friedmann a collection of problems in higher mathematics. Upon graduation from the University, taught at the Institute of Railway Engineering, at the Mining Institute in St. Petersburg (Petrograd), and from 1919 to 1922 at Simferopol (Tavrichesky) University. From 1922 till the end of his life, Smirnov's activity was connected with Leningrad University where he headed several departments and the Institute of Mathematics and Mechanics. Author of a widely used and recognized five-volume course of higher mathematics. Leader of the Leningrad mathematical school. Full Member of the USSR Academy of Sciences; Hero of Socialist Labor. Died in Leningrad on February 11, 1974.

Vladimir I. Smirnov.

Such is the short note about Vladimir Smirnov, in the format chosen for members of the "mathematical academy." We would like to add to this brief information that V. I. Smirnov was not only an outstanding mathematician and a famous historian of science, but also a person of exceptional nobility, benevolence and culture. All these qualities left a lasting impression even on those who seldom had occasion to meet this remarkable man in person, still more on his pupils and associates. Their love and repect for their teacher's memory were reflected in a three-day scientific conference which was held in Leningrad in June 1987 and was dedicated to the centenary of the scientist's birth. It is to be hoped that reminiscences about Vladimir Smirnov will be published and will thus become available to many thousands of readers.

Yakov Davidovich Tamarkin

Born on June 28, 1889 in Chernigov, into the family of a medical doctor. Studied in the same class as Friedmann in the 2nd St. Petersburg Gymnasium, which he left with a gold medal. After leaving school, he entered the mathematical section of the Physics and Mathematics Faculty of the university, and was retained to prepare for professor's work. Must have been V. A. Steklov's favorite pupil. In 1914, published *A Course in Analysis* in St. Petersburg. In July 1917, presented a Master's thesis at Petrograd University (N. M. Gunter and V. A. Steklov were the

Yakov A. Shokhat.

examiners). Friedmann's co-author in several works; published one work together with Steklov; co-author with Smirnov of the first volume of *A Course in Higher Mathematics*. Besides Petrograd University, he worked at the University of Perm (1919–20), at the Naval Academy, and at the Electrotechnical Institute. Emigrated to the USA in 1925, where he taught at Dartmouth College and Brown University. Published several books in the USA, one of them as Shokhat's co-author. A member of the American Academy of Arts and Sciences. Died in the USA in 1945.

Yakov Abramovich Shokhat

Born in the village of Roguznya (Kobrin district, Grodno province) on November 5, 1886. In 1920 left the Brest Gymnasium with a gold medal. Entered the mathematical section of the Physics and Mathematics Faculty of St. Petersburg University in 1906, graduated from it in 1910 and was retained there to prepare for professor's work. "A capable and hard-working person," as V. A. Steklov called him. Passed his Master's examinations in 1915. Taught first at the School of Commerce of the Institute of Forestry (Petrograd), then, in 1918–20, at the Mining Institute of Ekaterinburg (Sverdlovsk), and on return to Petrograd in 1920 at the Naval Academy and the Pedagogical and Polytechnical Institutes. In 1923 emigrated to Poland, where his parents lived, and from Poland to the USA. There he taught at the Universities of Michigan and California. The author of a number of monographs (one of them written jointly with Tamarkin) published in the USA, France and Germany. Died in the USA in 1944.

Teaching, research, Master's examinations

As we have seen, while studying for his Master's degree Friedmann studied many areas of mathematics and their applications. Besides this, he taught at the Mining Institute in close cooperation with N. M. Krylov and the Institute of Railway Engineering (where he cooperated most closely with N. M. Gunter). This teaching work was incidentally used by the university administration as a pretext to try and deprive him of his "graduate" stipend – and again Steklov's intervention helped remove this misunderstanding. It should be noted here that at the Institute of Railway Engineering, lectures on aeronautics were started in 1907; the first such course was given by N. A. Rynin. It is most probable that Friedmann was involved in this work at the institute. At any rate, in 1911, he published in two issues of *Problems of Physics* a brilliant popular article "On the theory of the airplane", in which, among other things, he summarized the content of fundamental articles by N. Ye. Zhukovsky and S. A. Chaplygin, as well as Professor van der Flit of the St. Petersburg Polytechnical Institute. Alexander Friedmann, as was always the case with him, showed himself in this article (which did not appear to be particularly ambitious and was written, most likely, to make a living) to be a well-grounded scientist, who had studied the background of the problem and, of course, knew all the contemporary publications. In this respect he was well ahead of Einstein who in 1915, i.e. four years later, got interested in these problems (in particular, the motion of surfaces in the air, which had been dealt with in detail by Friedmann), but "discovered" things already known, as was pointed out to the author of the theory of relativity in comments on his publication. Concerning his mistake, Einstein wrote in 1954: "This is what can happen to someone who thinks much, but reads little."[16]

During this period,, Friedmann also began at the Pavlovsk Aerological Observatory and even moved to Pavlovsk (this period is dealt with in more detail in Chapter 4). We shall only mention here that when he went to Pavlovsk he was already a married man: back in June 1911, in keeping

---

[16] See more about this in the article by V. Ya. Frenkel and B. Ye. Yavelov in *The Einstein Memorial Collection of 1980–1981*, Nauka Publishers, Moscow, 1985, pp. 37–48 (in Russian).

Friedmann among his friends. First row from left to right: A. F. Gavrilov, A. A. Friedmann, G. G. Weichardt, M. F. Petelin. Second row from left to right: Ye. P. Friedmann, Ya. A. Shokhat, Ye. G. Tamarkina-Weichardt, Ya. D. Tamarkin. Third row: V. I. Smirnov.

with the regulations of the time, he had addressed a letter to the Rector of St. Petersburg University, in which he "begged to be allowed to marry the spinster Ekaterina Dorofeyeva." Permission was given and a month later the young people were married in a church in the Petersburg *storona* (district) of St. Petersburg. Tamarkin found it necessary to inform Steklov about this event (on July 13, 1911): "The marriage of Alexander Alexandrovich was as unexpected to me as it was to you. His wife is quite a good-looking woman, although slightly older than he is. So far, I can say that the marriage has had only a positive effect on Alexander; it has reduced his habitual nervousness, made him calmer, and in no way hampered our studies, which have continued without interruption five times a week."

Ekaterina Petrovna Dorofeyeva, who graduated from some courses (probably the Bestuzhev Women's Courses) with the title of "governess," was a very quiet woman and had a considerable and beneficial effect on her husband, which was noted by Steklov, whose home Friedmann often visited with his wife from 1912. Many years later (or was it really so many?) in 1925, Steklov applied for a pension for Ekaterina Friedmann and highly commended the assistance that she had given to Alexander Friedmann, working on translations of his articles, reading proofs, etc.

At the end of 1912, having thoroughly prepared himself (of course,

together with Tamarkin), Friedmann began to take examinations for the
degree of "Master of Pure and Applied Mathematics." Let us stress "and
applied." It means that he was already thinking about applying mathematics to various specific problems (in physics, mechanics, meteorology, as
we can assume). The first examination was "in mechanics," as is recorded
in the minutes of the Physics and Mathematics Faculty of September 7,
1912. Friedmann was given the following questions: (1) On the motion of
a solid body possessing an immovable point on the axis of symmetry (the
case of Lagrange and Poisson); (2) On the motion of a solid body in a fluid
if there is no external force applied to the body, if the body has a surface
of rotation around its axis, and if its mass has the same axis of symmetry.

The report ends with a laconic entry: "The answer was considered
satisfactory." It sounds rather modest, but, according to the existing
regulations, there were no other marks to evaluate the answers of the
would-be Masters of Science. The report was signed by D. K. Bobylev
(mechanics), V. A. Steklov (mathematics), and I. I. Ivanov.

The second question was probably formulated by Steklov: his own
Master's thesis was on the motion of solid bodies in a fluid.[17]

There is a piece of evidence showing the importance Steklov attached to
the studies of his students. He writes in a brief note in his diary of
December 7, 1912: "This evening Tamarkin, Friedmann and Bulygin
took Master's examinations. Their answers will, of course, be quite good.
There were Bobylev, myself, I. I. Ivanov and, at the beginning, Selivanov
and A. A. Ivanov. They [Tamarkin, Friedmann, Bulygin – V. F.] were, of
course, in raptures and are celebrating today their first examination."

And each time Steklov attended examinations, he commented on his
students' answers. As for mathematics, Friedmann took this examination
as mathematicians sometimes take integrals – by parts: on 13th and 29th
of March, 24th of April and 1st of November, 1913. Steklov wrote in his
diary on March 13: "At 12 o'clock today Selivanov examined (in finite
differences) Friedmann, Tamarkin, Smirnov, [Shokhat] and Bulygin.
Everybody was finished within an hour (even less). On the 22nd, three of
them will be examined by Markov, and on the 29th the remaining two
(Friedmann and Smirnov)." Now, we know that each item of the examination had a corresponding chief examiner. In this case, Selivanov asked
Friedmann to give a method of solving linear differential equations with
constant coefficients and without the last term (in other words, homogeneous equations), and "linear equations with periodic coefficients"

---

[17] This was incidentally the topic of P. S. Ehrenfest's doctoral thesis done in 1914 in Vienna
under L. Boltzmann's supervision.

(Fuch's theorem).[18] Friedmann gave an answer to this question to Yu. V. Sokhotsky, and on "Euler integrals and Fourier's theorem" to (of course) V. A. Steklov. On March 29, Steklov writes in his diary: "At 2.30 p.m. Friedmann and Smirnov were examined in mathematics by Andrei Markov. They naturally got through. Markov finished very soon – within half an hour." One should remember that A. A. Markov was an exacting examiner who made many physicists and mathematicians suffer. (According to reminiscences by A. F. Ioffe, P. S. Ehrenfest and others, physicists taking examinations in mathematics in the late 19th and early 20th century suffered most from Korkin and Markov. They had no clear program, and when asked about the scope of questions that were to be prepared and what constituted the subject matter of mathematics, Korkin once answered: "Mathematics is what Markov and I do.") Friedmann's next examination was on April 24, and Steklov wrote: "Tamarkin, Friedmann, Smirnov and Shokhat gave very good answers."

But sometimes, Vladimir Steklov gave more reserved assessments. Thus, at the final examination in mathematics which took place in the autumn of 1913, in November, Friedmann and Bulygin were asked by Sokhotsky about the theory of gamma functions and Fourier's integrals. This time they "did not give good answers, got confused," as Steklov wrote, although they passed the examinations. There were examinations which were less important for mathematicians, in physics and meteorology, in particular. Friedmann passed all these examinations successfully, winning the right to be called a "Master's degree student." There was one important step from a Master's degree student to a Master's degree: the thesis. We may as well say now that it took Friedmann many years to take this step, and Friedmann defended his Master's thesis in 1922, after being a professor at Perm university and (by 1922) a professor at the Institute of Railway Engineering, the Polytechnical Institute, and the Naval Academy.

[18] Schrödinger's equation describing the movement of an electron in a periodic field belongs to this class.

4

In search of a way

Master's studies

Preparations for Master's examinations left almost no time for research: the theoretical courses which were to be studied for the examinations were quite extensive, and the examination requirements were extremely challenging. In July 1911, in the letter in which Friedmann informed his professor about his forthcoming marriage, he also gave a brief "graduate student's" report: "Our studies with Yak. Dav.[1] seem to be going quite well. They have naturally been confined to reading the courses you recommended and articles for the Master's examinations. We are through with hydrodynamics and are going on to study the theory of electricity." Therefore, during his Master's studies Friedmann published relatively few papers. In one article – 'On finding particular solutions of the Laplace equation", which appeared in 1911 in the *Bulletin of the Kharkov Mathematical Society* – he solved, as any Master's degree student was supposed to do, the problem set in his teacher's doctoral thesis: find all orthogonal coordinates in which Laplace's three-dimensional equation admits of partial separation of the variables. In solving this problem, according to V. A. Steklov's testimonial, Friedmann 'displayed the required ingenuity and a good knowledge of analysis." Another paper on the theory of partial differential equations, "On finding isodynamic surfaces," came out in 1912. Friedmann learned about the problem of isodynamic surfaces from A. E. H. Love's textbook on the theory of elasticity. In Lamé's definition, isodynamic surfaces are surfaces in an elastic body, on which there are no oblique tensions, and each normal tension depends on only one coordinate. In an article published in the *Bulletin of the Paris Academy of Sciences*, Friedmann announces that an exhaustive

---

[1] Yakov Davidovich Tamarkin.

classification of such surfaces has been worked out for two cases – when none of the tensions, and when at least one of the tensions, is constant.

Friedmann also tested himself in applied problems. Jointly with V. I. Smirnov he discusses the problem of condenser discharge for resistance linearly changing with time; in the article "On the theory of airplanes" he gives an approximate theory of movement of curved surfaces in the air. Jointly with M. F. Petelin he solved the aerodynamic problem of V. Bjerknes on the interaction of two spheres placed in an ideal fluid. Bjerknes considered oscillations of the surfaces of the spheres, and observed that under certain conditions the pulsating spheres interact on average as bodies attracted to each other according to Newton's law (Coulomb's law to be exact, because the problem admits of both attraction and repulsion). Petelin and Friedmann set themselves the task of finding oscillation patterns for spherical surfaces such that the interaction is proportional to the inverse-square distance at each moment of time. They also reduced the problem to a system of two ordinary non-linear equations of the second degree and found its first integrals.

Some of the papers begun at that period were left unfinished, e.g. those on problems in the theory of elasticity, in which Friedmann attempts a solution using the method of fundamental functions, but discovers an inaccuracy in reasoning. Several years later, he discovered the same mistake in a paper, apparently by a foreign author, on the application of the theory of elasticity to seismology, which he was asked to look at by the Director of the Main Physical Observatory, B. B. Golitsyn.

Friedmann had to do some mathematical donkey work too. In 1910, waiting for the completion of formalities necessary for getting a stipend, he applied, in search of an income, to the famous ship-builder A. N. Krylov, at that time a professor at the Naval Academy. Alexei Krylov referred Friedmann to another outstanding ship-building engineer, Ivan Grigorievich Bubnov, who was a professor at the same academy and also taught at the Polytechnical Institute. Bubnov had some calculation work for Friedmann. His task was to measure the sag of an elastic plate of complex shape using the method of another professor, N. M. Gersevanov. This work was quite timely for Friedmann, and he set out to do it with enthusiasm, starting with a criticism of the suggested method. We learn from his letter to Steklov that several days later he wrote to Bubnov: "I am not getting on well with calculations using Gersevanov's method: Gersevanov did not prove either the convergence of the series obtained or even that its coefficients can be obtained from his equation," adding that an attempt to apply the method to a simpler task showed its invalidity.

Friedmann suggested applying the method recently suggested by Ritz, with some improvements made by Tamarkin. Bubnov agreed, and Friedmann set out to make fairly elaborate calculations. Readers familiar with applied mathematics and mechanics may ask why Bubnov did not recommend the Bubnov–Galerkin method, which is a currently widespread approximate method directly applicable to the task in question. But Galerkin's investigations on this question were started in 1915, and Bubnov's article in which this method was applied to a non-linear problem in the theory of plates appeared in 1916. On the other hand, as far as can be judged from Friedmann's letter to Steklov of August 3, 1910, the calculation technique Friedmann used is equivalent to the technique outlined in the Bubnov–Galerkin method. So, it is quite possible that Tamarkin was the one who invented this technique, and that Friedmann was the first to use it in practice.

Apart from the obligatory program of Master's examinations, Friedmann studied new books in mathematics. A. F. Gavrilov recalls: "Alexander Alexandrovich was among the first people at the University to understand the theoretical and practical value of integral equations, which did not enjoy much respect among the older generation of mathematicians." The discussion between the younger and older generations of mathematicians is reflected in a letter from Ensign Friedmann to Professor Steklov (July 25, 1915): "Concerning the same question [the solution of an integral equation of Volterra type – E. T.], I had to get acquainted with Volterra's works on the function of a line or of a function. How do you feel about these ideas? It seems to me that this is no longer decadence, and that in many cases there are good methods here. I may think that way, though, because of 'long starvation.'"

The Main Physical Observatory

Thus, Friedmann was trying to find his way in science in his first post-university years, thoroughly studying recommended textbooks and articles, making his first independent investigations, coming over from cumbersome "hack" calculations to reading the latest mathematical literature, considered decadent by the professors. Let us refer again to A. F. Gavrilov's reminiscences:

A. A. Friedmann had real talent for mathematics, but he was not satisfied solely with the study of the mathematical world of numbers, space and functional relations in them. The world studied by theoretical and mathematical physics was not sufficient for him either. His ideal was to observe the real world and create a mathematical apparatus which would allow us to formulate the laws of physics with adequate generalization and depth, and then, without observation, to predict new laws.

He chose the study of the atmosphere which, according to him, is an immense laboratory for which mathematics has prepared a method of study in the form of potential theory.

The last statement, if taken literally, shows the enthusiasm of a youth rather than an exact assessment by a representative of university science, yet the choice Friedmann made in 1915 was absolutely correct, although he could not at the time judge himself as definitely as Gavrilov characterized him in 1959. In early 1913, on Steklov's recommendation, Alexander Friedmann got acquainted with M. A. Rykachev, the then director of the Main Physical Observatory, and B. B. Golitsyn, who was to replace Rykachev several months later. Steklov wrote in his diary on January 13, 1913: "Yesterday I gave A. A. Friedmann a letter to M. A. Rykachev. I'm asking him to give Friedmann, if possible, a job either at the physical section of the Observatory or, better still, at the aerological [section]. Let's see what comes of it." It so happened that Friedmann got a position as a physicist in the Aerological Observatory in Pavlovsk, which had been part of the MPO since February 1913. It was a serious step which significantly changed the nature of Friedmann's activities. He was getting off the "straight track" – a thesis, professorship at the University, research work within the school of mathematical physics formed around Steklov – the track along which his colleagues (Smirnov, Tamarkin and others) were moving with much success. If Friedmann had been asked at the time to assess his action, with his usual self-irony he would have probably attributed it to his "changing character" or worldly considerations. But the intensity with which he began to study the area which was new for him, and a rapidly increased scientific productivity, testified that the choice Alexander Friedmann had made was useful both for him and meteorological science. He moved to Pavlovsk and became utterly engrossed in studying the methods of aerological observations, and dynamic and synoptic meteorology.

The first task assigned to Friedmann at the Observatory was the processing of observations done with kite meteorographs. As with the task Bubnov had given him three years before, Friedmann could not limit himself to calculations which, in Gunter's phrase, had been "conventionally done for decades." Friedmann soon wrote instructions which introduced simplifications and time-saving techniques into observation processing.

The Main Physical (later Geophysical) Observatory was an institution of the Academy of Sciences. Its director was elected by the general meeting of the Academy, and was himself required by its regulations to be a Full Member of the Academy, while administratively and financially the Observatory was subordinated to the Ministry of Education. The Observatory was set up in the mid 19th century. In his article on the history of the Observatory M. I. Budyko writes that the idea of "instituting meteorological observations across Russia" was first put forward by V. N. Karazin, a famous public figure, "Marquis Posa" to the Russian "Don Carlos" in the first liberal years of the reign of Alexander I. The history of his relations with the Emperor (broken by a court intrigue) and his many scientific and social projects are vividly described in A. I. Hertzen's article "Emperor Alexander I and V. N. Karazin."[2] The proposal to set up a central meteorological institution was formulated by Karazin in 1808, four years after the suspicious Tsar had been "offended" by him. It did not succeed. In the 1830s, this idea was pushed by Academician A. Ya. Kupfer, but it was only in 1849 that the Main Physical Observatory was set up. It became the second major scientific institution of its kind in Russia after the Main Astronomical Observatory in Pulkovo.

A major stage in the history of the Observatory is linked with the name of the Swiss meteorologist Heinrich Wild, who was Director of the observatory in 1868–96. During his tenure, the Aerological and Magnetic Observatories were set up in Pavlovsk, as well as several regional observatories and hundreds of meteorological stations. During his short time as Director of the Observatory, in 1916, A. N. Krylov got acquainted with the order of things H. Wild had introduced, and was rather critical of them; but he gave credit to Wild the scientist, although not without hidden irony: "It appears that H. Wild was a man of great learning and unusual diligence. He left a lot of calculation work behind him; it seems he wanted to extract general laws from a mass of observations. He applied harmonic analysis to these observations, with both yearly and daily periods for a given place, and tried to apply expansions in terms of spherical functions for different places, as Gauss did for terrestrial magnetism. He built original and most effective magnetic instruments, stationary ones for the Magnetic Observatory in Pavlovsk, and portable one for magnetic survey." In 1896, Mikhail Alexandrovich Rykachev was elected Director of the Observatory. As Krylov described him, "M. A.

---

[2] *Pole Star* of 1862. Seventh book (in two parts). Nauka Publishers, Moscow, part 2, 8–56 (in Russian).

Rykachev belonged to the school of H. Wild whose assistant he had been for many years, and he continued to manage things the way Wild did." Indeed, by the time of his election as Director, M. A. Rykachev had served in the observatory 29 years already. In the initial years (1868–73) he made several flights in a balloon to study the free atmosphere. Having become Director, he continued to develop the network of meteorological stations and the weather forecast service. Research was started in synoptic and agricultural meteorology, and in actinometry. The Main Physical Observatory acquired international prestige, expressed by the election of Wild as President of the International Commission on the First Polar Year, and of Rykachev as Chairman of the First International Aeronautical Congress.

In 1913, Academician M. A. Rykachev, Admiral of the Fleet, asked to be relieved of his post. Academician B. B. Golitsyn was elected Director. He was an outstanding geophysicist, but his scientific interests lay not in the field of meteorology, but in the field of seismometry and seismology, for the founding of which he is rightly given credit. It is hard to say whether Golitsyn had any influence on the work of the meteorological and synoptic departments, but the fact of inviting Friedmann to work at the observatory shows that Golitsyn noticed that the observatory was lagging behind in the field of theoretical meteorology.

A research student with Bjerknes

The leading school of theoretical meteorology at the time was that of the Norwegian scientist Vilhelm Bjerknes. In those years Bjerknes was lecturing at Leipzig University and headed the Univerity's Geophysical Institute where his talented disciples – Hesselberg, Sverdrup and Holtsmark – worked. In 1904, Bjerknes proposed a program of weather forecasting based on the equations of atmospheric hydrodynamics and thermodynamics, plus a fairly precise knowledge of the initial conditions of the atmosphere at a given moment of time. Bjerknes hoped that one day theoretical meteorology would attain the level of celestial mechanics and would be able "to determine movements of air masses in the Earth's atmosphere and changes in the state of these air masses produced by their motion," as celestial mechanics determines the coordinates of astronomical objects. It is interesting to compare this thought of Bjerknes's with M. V. Lomonosov's (1711–65) idea that the problem of weather forecasting might in principle be solved in the near future, provided science had "a true theory of motion of fluid bodies near the globe, i.e. of water and air." Although not as soon as Lomonosov had hoped, but by the beginning of the 20th century, meteorologists had such a "true theory" – a closed system of differential equations relating to the air's velocity, density, pressure, temperature and humidity. It was therefore possible to attempt the solution of the system of equations, provided one knew the initial data, boundary conditions, and data on heat input. Scientists were yet to find out that this system of equations has a property significantly limiting the possibility of forecasting – instability, reflecting the instability of the real atmosphere with regard to minor disturbances. In the 60s of this century Edward Lorenz would call this property the "butterfly effect," and formulate it as a question-statement: "Can a butterfly flying in Peru cause a tornado in Iowa?" At the beginning of the century, it was clear to Bjerknes that there was no possibility of obtaining an analytical solution of this complex system of equations. He pinned great hopes on graphical methods. These hopes proved to be illusory. The task of weather forecasting on the basis of a system of hydrodynamic equations had to wait for the advent of electronic computers. So far Bjerknes' graphical methods had been applied to the more modest task of processing aerological observations. The technique was described in his two-volume course in dynamic meteorology which Friedmann had studied back in Pavlovsk. The first volume dealt with the problem of the statics of the atmosphere and gave the techniques of making pressure maps for different altitudes. The second volume tackled the task of drawing lines of air current flow. Not long before Friedmann's arrival in Leipzig, Bjerknes started to publish *Synoptic Maps* – pamphlets containing distribution maps of meteorological elements at different altitudes according to measurements made on international aerological days. These "notebooks" immediately became popular with the meteorological community. Every researcher who came to Bjerknes for short-term studies began by processing aerological observations. So did Friedmann, who processed the data from the observations of one international day. The results were published in a regular issue of *Synoptic Maps* without the author's name, in keeping with the established tradition.

The frontal method of weather forecasting was adopted by Soviet meteorologists as well, albeit after Friedmann's death, in the 1930s. In Leipzig Friedmann tackled the problem relating rather to the future numerical forecasting method. It was a manifestation of his constant desire to "eliminate things," i.e. to remove insignificant terms in the equations of hydrodynamics so as to make them solvable. Friedmann was grappling with this task together with T. Hesselberg. In Leipzig Friedmann made the acquaintance of a number of outstanding meteorologists who later, after World War I and the Civil War, helped him to re-establish scientific ties of Soviet meteorologists with foreign colleagues, and Hesselberg and he became close friends. They wrote a joint paper "The order of magnitude of meteorological elements in the space and time derivatives." The authors' purpose was to assess the relative value of individual items in the hydrodynamic equations of a compressible fluid. To assess the derivatives numerically, they used an approximate representation through ratios of finite differences. Like all geophysicists of the time, Hesselberg and Friedmann worked with measurable quantities using the MTS system of units – meter, tonne, second. To calculate the finite differences of meteorological elements it was necessary to set the finite differences ("steps") of coordinates and of time. Considering that the horizontal change of meteorological elements are 100–1,000 times smaller than the vertical ones, the authors chose the following values: the step in length in the horizontal direction $\Delta s = 10$ km, the step in length in vertical direction $\Delta z = 0.1$ km, the step in time $\Delta t = 3,600$ sec. Determining, from observational data, the horizontal and vertical components of velocity, as well as the pressure, specific volume and temperature, and calculating approximate values for their derivatives with respect to horizontal and vertical coordinates and time, Hesselberg and Friedmann constructed tables containing the orders of magnitude of all the terms in the hydrodynamic equations. The tables included typical values of the terms, which were determined using a great deal of statistical material. Then a procedure for simplifying the differential equations was given, in the case when some terms in the equation were significantly smaller than the others. It was shown in particular that the total time derivatives of velocity components cannot be simplified, and the total pressure derivatives can.

The article by Hesselberg and Friedmann published by the Geophysical Institute of Leipzig University[3] after World War I had started was noted by theoretical meteorologists and later became their desktop guide. M. I. Yudin, the head of the dynamic meteorology department of the Main Geophysical Observatory named for A. I. Voeikov, having slightly modernized the Friedmann–Hesselberg method, resorted to it even in the 1950s. However, their paper as seen by today's aerodynamics specialists produces a mixed impression. It was, undoubtedly, a step in the right direction – towards a system of equations precise enough for forecasting but not too complicated for solution. But the authors did not construct

3 *Veröffentlichungen*, Serie 2 (Spezialarbeiten), Heft 6.

such a system (Friedmann's pupil I. A. Kibel was to produce it almost 30 years later). If Hesselberg and Friedmann had applied their method not to separate groups of items, but to all the terms of the equations, they could have discovered the so-called geostrophic model of the atmosphere, which takes into account in equations of motion only the Coriolis force and the pressure gradient. They were done disservice by the standardization of the system of geophysical units, which helped them to avoid the necessity of using dimensionless values, but also deprived them of the possibility of introducing the dimensionless parameters (criteria of similarity) which played such an important role in hydrodynamics in later decades. "In some questions of science there is a clear tendency towards standardization and the administrative introduction of the universal system of units," writes Academician L. I. Sedov.[4] "In a number of cases it is apparently very convenient and useful, yet to attach the universal system of units to certain physical constants or conditions is in many cases artificial. Conversely, the very possibility of using arbitrary units of measurement, and the independence of the laws under investigation of the choice of the system of units, can lead to useful conclusions." This view, now commonly accepted, was yet to establish itself in hydromechanics. Today the researcher always starts work by determining the scale of the phenomenon under investigation, and through a virtually instantaneous assessment of dimensionless parameters (named for major scientists of the past, such as the Reynolds, Mach and Euler numbers) decides which theoretical model is suitable for describing it – thus answering, for this particular case, Friedmann's question "What can be eliminated?" For a researcher of Friedmann's type the theory of dimensionality and similarity would have been a useful instrument, as is confirmed by the way it was used by his pupils I. A. Kibel and N. Ye. Kochin. Friedmann himself unfortunately did not have command of this instrument – simple, but indispensable in studying such a complex ("multidimensional") object as the atmosphere – and this made not only the Leipzig paper of 1914, but also his later works, somewhat incomplete.

In the above-quoted textbook Sedov draws an interesting analogy between the choice of the system of measurement units and the choice of the system of coordinates and the system of references: "One can, of course, fix a quite definite universal system of reference, and study all phenomena only within this system. Yet, the very possibility of using different systems of reference, and the possibility of applying special

[4] L. I. Sedov, *Mechanics of Continuous Media*, Vol. 1, Nauka Publishers, Moscow, 1970, p. 399 (in Russian).

characteristic systems of reference to different tasks, is the basis of a fruitful method of physical research." Nine years after the Leipzig paper the famous words will be spoken: "We will soon understand the principle of relativity: Friedmann has set out to study Weyl." There is no doubt that, just as effectively as he tackled the theory of relativity, Friedmann would have got an understanding of the invariance of physical laws relative to a group of similarity transformations, but he was not destined to accomplish this.

Having finished his work with Hesselberg, Friedmann left Leipzig. In the summer he took part in the development and preparation of aerological observations which were to be conducted during an eclipse (in August 1914), and for this he took several flights in airships. Then, on August 1, 1914, the World War broke out. Leipzig and Pavlovsk found themselves on different sides of the Russian–German front, and Alexander Friedmann's career took a new and drastic turn.

5

War years

The first months

When the war began, Vladimir Steklov was in Britain with his wife. They did not manage to return to Russia the usual way – via Germany – and set off for home by a long way: by sea to Norway, then via Sweden and Finland with a stop at Viborg. In the evening of August 5 (Old Style), the Steklovs arrived at the Finland Station in St. Petersburg and suddenly stumbled across Friedmann. The encounter proved to be quite timely, because Steklov had no money at all with him, so Friedmann gave him some and helped to hire a cab. The next day Steklov wrote in his diary: "Friedmann turned up unexpectedly . . . Had volunteered to join up and serve in an aviation company; sent by the Main Physical Observatory. He helped fix the lighting for the fuses had blown, otherwise we would have been sitting in the dark. So he put things right, Olga had come across some spare fuses. Had a lot of tea with rolls bought at Viborg, Friedmann left at 1 a.m."

Friedmann writes in his autobiography that "in order to introduce aerological observations into aviation practice and thus, on the one hand, provide a service to aviation, and, on the other hand, increase the number of aerological stations, he joined, with the permission and approval of B. B. Golitsyn, Director of the Observatory, the volunteer aviation detachment; in which he worked, first on the northern front, near the towns of Osovets and Lyk, and later on other fronts, to organize aerological observations and aeronautical services in general."

Thus began Friedmann's war odyssey, which lasted for some three years. Information about his service in the Army has been preserved: Academicians Golitsyn and Steklov kept his letters from the war fronts. These letters, sincere and lively, showing him as an exceptionally brave man extremely dedicated to science, are so good that it would be unjust

68

merely to paraphrase them. We believe that the heroes of biographical books should always be given a chance to speak for themselves – if the opportunity exists. We have such an opportunity, and things are made easier by the fact that the above-mentioned letters are included in Friedmann's *Selected Works*. We believe that some of them could be interesting to war historians – they give a vivid picture of the events of the summer of 1915, a bitter period for the Russian Army, as seen by a person thrown into this maelstrom.

Friedmann was fairly often sent to Petrograd, and there he invariably visited V. A. Steklov.

Vladimir Steklov wrote on September 2, 1914: "Yeterday morning A. A. Friedmann came. He has not yet left and now will probably stay to defend the city with his airship company." On the same day he wrote in his diary: "St. Petersburg has been renamed Petrograd by Imperial Order. Such trifles are all our tyrants can do – religious processions and extermination of the Russian people by all possible means. Bastards! Well, just you wait. They will get it hot one day!"

November 2, 1914: "On the way to the University I met Friedmann! He is being sent by Grand Duke Alexander Mikhailovich to different regiments to organize air reconnaissance. Nominated as a candidate to be awarded the Order of St. George (fourth class) for successful reconnaissance. He had a lot to say. We had lunch together. He says that the army's morale is high, although we naturally wished for the war to end in victory soon. He will stay here probably for about two weeks and will then go to Lvov . . . He is fairly satisfied with the general situation and his activities."

November 23, 1914. "This evening we were visited by Friedmann, Petelin, Shokhat, Tamarkin, Smirnov, Gavrilov, Gunter, Bulygin. Friedmann had got the Order of St. George for reconnaissance flights (in an airplane). Gavrilov is already wearing a conductor's uniform. Friedmann will soon go back to the war, to East Prussia. When Gavrilov will be sent, nobody knows. They left at 1.30 a.m."

Friedmann left Petrograd for Lvov on December 4, having visited Vladimir Steklov together with Petelin on December 3. The day before, Steklov had got several offprints from T. Levi-Civita, and Friedmann asked for them for "temporary use." Steklov writes that Friedmann "was not in very high spirits. Well, God is not without grace, and a Cossack is not without luck! We shall see!"

Below are excerpts from Alexander Friedmann's wartime letters. Most of them date from May–July 1915. They were written during a difficult

Vladimir A. Steklov.

period, when the situation in Galicia was very hard for the Russian Army. Having recovered after their defeats in August–September 1914 on the south-western front, the Austrian–German troops broke the front in the area of the 3rd Russian Army on May 2–5, 1915, capitalizing on their superiority in heavy artillery. On June 3, Przemyśl fell, on June 22 Lvov, and the Russian Army left Galicia. We have omitted those passages in the letters which show that Friedmann used every opportunity to continue his work, obtaining new results in atmospheric physics and mathematics and discussing them with his teachers. Let us note that after each letter from his pupil Steklov recorded it in his diary, sometimes giving the contents in detail. Is this not another piece of evidence showing his attachment to Alexander Friedmann?

Letters from the Front

February 5, 1915

Dear and highly esteemed Vladimir Andreyevich [Steklov],

Today I received your postcard and would like to express my heartfelt gratitude to you and Olga Nikolaevna for remembering me and for the gift which I have not yet received, but will probably get soon. My life is fairly even, except such accidents as a shrapnel explosion twenty feet away, the explosion of an Austrian bomb within half a foot, which turned out almost happily, and falling down on my face and head, which resulted in a ruptured upper lip and headaches. But one gets used to all this, of course, particularly seeing things all around which are a thousand times more awful.

A good Austrian airplane was seized recently; I talked much with the captured pilot; the fellow was rather sly, so from the talk with him one could get the impression that everything is beautiful in Austria, there are enough troops and ammunition, and that the war, in his view, will end with their victory. This is, of course, nonsense, but that the war may become protracted is fairly probable.

As for me, on the completion of the aerological mission I'm thinking of learning how to fly; this is no longer very dangerous and can be successfully used in meteorology and particularly in synoptic observations.

I've been giving a lot of attention recently to the theory of bomb-dropping, which is one of the tasks given by the Grand Duke [Grand Duke Alexander Mikhailovich, the uncle of Nicholas II, "Most August Commander of Aviation and Aeronautics in the Army in the Field" – V. F.]. The question is reduced to the following equations:

$$\frac{du}{dt} = -au\sqrt{(u^2 + v^2)}, \qquad \frac{dv}{dt} = -g - av\sqrt{(u^2 + v^2)};$$

at $t = 0$, $u = c$ (about 20 to 40 m/sec) and $v = 0$; u, v are the components of the bomb velocity along the coordinate axes, g is acceleration due to gravity, and a is a parameter characterizing the shape and weight of the bomb. The bombs used can be divided into two classes: those of the first class have a very small a, and the others have a close to 1. In solving the problem in the first case, I expanded u and v in powers of a, and for practical purposes it turned out to be enough to confine myself to the lowest powers of a; for the second class of bomb I used a method which is mathematically illegitimate, assuming approximately

$$\sqrt{(u^2 + v^2)} = v\sqrt{[1 + (u/v)^2]} \approx v.$$

The results are somehow very close to reality, so they have been used so far in practice (the arsenal in the fortress has already been destroyed) [Przemyśl is meant – V. F.], but not just lack of rigor, but also the mathematical ambiguity of this rather dubious method keeps me on the lookout for other methods, yet I haven't found anything.

It's difficult to study due to lack of books; I am now reading Wien's *Hydrodynamics* and rereading Zhukovsky's *Kinematics of a Fluid Body*.

Steklov's diary, February 13, 1915. "I have sent a letter to the army in the field to Friedmann. I show in it that his equations can be integrated in quadrature, and that the principal values are conveniently expanded in powers of $1/c$, where c is the initial bomb velocity along the y-axis (parallel to the horizon). He won't take my advice to keep off bombs and get more into calculations: it would be more productive too: he should not fly either, others will do that better than him and there is nobody to perform calculations! Let him do what he can do best, i.e. calculations."

March 5. "Olga got a letter from A. A. Friedmann from Przemyśl; he thanks her for the clothes, the warm clothes proved to be particularly useful. It's very cold there, especially during flights."

February 28, 1915

My dearest Vladimir Andreyevich [Steklov],

I am very grateful to you for remembering me and for the gift. As I have already written to Olga Nikolaevna, the warm clothes turned out to be very useful indeed, because it is now extremely cold. I will also need them for next winter since there is no end to the hostilities in sight . . .

Many thanks for the equation. I stopped halfway, having obtained the equation for dx/dt and dy/dt ($x = \sqrt{(u^2 + v^2)}$, $y = v/u$), but I did not have the sense to express y in terms of x. Although the formulae obtained are rather long, a firm base makes it possible to tackle them and, proceeding from them, give rigorous expansions in series. I had been uneasy about the lack of rigor in my reasoning, and although the calculations agreed with practice fairly well, still this is not the way it should be, it does not feel as good as when you have expansions in which one can see what is eliminated and what is taken into account. Although approximate integration using Runge's method produces, of course, a quick result in terms of calculations, it also forces one to assume a as given, which is not always convenient, because one often has to determine a by the time taken for the bomb to drop from a definite altitude.

I have recently had a chance to verify my ideas during a flight over Przemyśl; the bombs turned out to be falling almost the way the theory predicts. To have conclusive proof of the theory I'm going to fly again in a few days. The bombs I drop (5 lb, 25 lb and 1 pood [40 lb] in weight) belong to the class in which a is very small, so I've been verifying the expansion of the solutions in terms of the parameter a.

It should be said that it is fairly difficult to do investigations in my circumstances, and all kinds of calculations concerning bombs and bomb-sights take a lot of time, therefore I have very little time for the theory of vortices in fluids with changing temperatures for which I feel a strong attachment . . .

Let us now interrupt the citation of that letter. Przemyśl (in Friedmann's letter coded by the initial P; we had to "decode" it) was a strongly fortified Austrian fortress in Western Galicia. It was blockaded by Russian troops in late September 1914, seized by them on March 22, 1915, and abandoned two months later. So Friedmann's flights over Przemyśl were made during its six-month blockade. It so happened that H. Ficker, a future professor of meteorology (whom Friedmann was destined to meet in peacetime in 1923) was in 1915 at Przemyśl, then occupied by Austrian troops. In Friedmann's obituary, V. A. Steklov relates that in 1925, on learning from Steklov about Friedmann's death, Ficker wrote back saying that the only bomb he saw hit the target in Przemyśl was dropped from the airplane Friedmann was in.

A natural question arises: how could Ficker know this? The answer is given by the quotation from Ficker's extremely sympathetic interview which he gave to a correspondent of the Leningrad *Krasnaya Gazeta* (Red Newspaper): "I met with Friedmann three times in Berlin. One day we had a talk and it turned out that not long before the fall of Przemyśl Professor Friedmann, being a Russian military pilot, dropped a very powerful bomb over my home there. When the bomb fell, being a German pilot, I was in my superior's office to receive my orders. I remember well that the bomb which Friedmann dropped was the only Russian bomb which hit the target at Przemyśl. When I first met Friedmann in Berlin we found out the exact time and place of our unusual and unfriendly acquaintance on the battlefield."

In an article of memoirs about Alexander Friedmann it is mentioned that when bombs hit targets at Przemyśl, German soldiers would say: "Friedmann is in the air today." It remains, of course, unclear how they learned about it. If, however, this is a legend, then it's a very characteristic one!

Now let's go back to Friedmann's letter to Steklov:

My life is fairly varied, mostly I'm busy organizing different kinds of observations and suchlike. The other day I was working on the compass deviation in the airplane and was several days struggling with the expansion of the results into trigonometric series.

Thank you, Vladimir Andreyevich, for your kindness. Looking forward to hearing from you again – every letter here is very, very precious. I think we will see each other soon, as I have accumulated a lot of material, so that I will have to go to Petrograd; I may come back already an officer – it's about time, as I have been in combat for eight months already.

Sincerely and respectfully yours,

A. Friedmann.

In April Friedmann found himself back in Petrograd already an officer; he spent about three weeks there. On April 15, he had lunch at Steklov's place. "He says that there is no end to the war," wrote Steklov. Three days later Friedmann and his wife were again at the Steklovs' place, together with his colleagues, staying as usual well after midnight. On April 29, the Steklovs gave a party for former students – the Friedmanns, Petelin, Smirnov, Tamarkin, Shokhat. There were also members of the older generation – D. A. Grave, Ya. V. Uspensky, the Sintsovs. Time flies in conversation: the guests left at 2.20 a.m. Before leaving for Lvov, Alexander Friedmann came to see his teacher again on May 5 and 7. They spoke about science as well: discussing Friedmann's aerodynamics papers.

The following letter is to Boris Borisovich Golitsyn:

May 20, 1915, Lvov

Dear Boris Borisovich,

I am sending you this letter by someone travelling in your direction. Since it will not go through the usual mail channels, I may tell you some details. At present our situation is satisfactory – we are advancing, and if you look at a map of the Front you will see that the Germans can easily get caught in a pocket in Galicia. But several days before May 20 the situation was different. One of the units of our army suffered a crushing defeat, thus, in one regiment only two officers and a few dozen non-commissioned officers survived. There was a panicky mood in Lvov: civil offices and army units hurried to move to the rear. There were rumors that it had been decided to retreat to Kiev or at any rate to evacuate Galicia; fortunately, this did not happen. The causes of our failures are twofold: on the one hand, heavy artillery in the German Army and lack of it in ours, on the other hand, the excellent organization and equipment in the German Army and lack of either in ours. The heavy artillery has a tremendous effect on morale – the actual damage from it is not great. But in modern warfare the morale of the infantry is all-important.

Life in Lvov is now absolutely quiet, the only entertainment is the enemy's airplanes, sometimes, though seldom, dropping "presents." It can be said that the airplanes do not produce any dispiriting effect, or indeed any serious effect on morale.

I go on with my duties and try to improve, but everything is ten times as slow as I would like it to be. There are tremendous obstacles, mainly because of personal clashes between different authorities. All this tells on my nerves and I have to say that I am sick and tired of all this, and it has cost me a lot of effort to keep my energy from falling below the permissible level. If there is anything that gives me strength now, it's my scientific work and thinking about the Observatory and my future work in it . . .

I will soon be leaving Lvov to inspect stations in detachments; there I will try to get the opinions of the commanders about aerological observations, and on the basis of these opinions I will compile a draft list, both of observation equipment to be assigned to aviation detachments, and of staff at the central stations in companies and the general inspectorate of stations. If you permit, I will send you this draft and I hope that you will kindly make your corrections.

To B. B. Golitsyn:

May 25, 1915
Army in the Field
M. Glembochek

Dear Boris Borisovich,

I am taking the opportunity to send you a letter by a reliable person. The military situation, as you probably know, is quite bad. They are already near Riga (Mitava is abandoned), Osovets will soon be seized or has already been abandoned; Warsaw must be, Novogeorgievsk must be, Ivanogorod may be, Lublin must be, Vladimir-Volynsky must be, and Kholm must be in their hands. We are holding only part of Galicia; here our front goes along the Zolotaya Lipa, the Dniester until Usupi, Biskune and then along a straight line southward to the Romanian border. The situation has become difficult: there is a clear superiority in the enemy's forces and so far we have had to retreat, straightening our line; our allies are somewhat slow to act; this is due to their being used to conducting the war in comfort in beautifully furnished trenches. If they are too slow, if they let the Germans make us retreat, with terrible bloodshed, and straighten out the front line, then, most likely, the Germans will greatly strengthen their positions and will move part of their troops to the west; it's hard to imagine what would then happen to our allies. The war in front may drag out for a very long time, thereby fulfilling my prediction that we will come back captains; well, I may not become a captain, but for sure I will soon be promoted to second lieutenant. I am telling you some general information, but I ask you to keep it secret, because this is information which should by no means be disclosed, although it's quite reliable. However, in five days, when this letter reaches you, this secret information will probably be given wide publicity and become a sad truth.

The character of aviation is now changing rapidly; the techniques of armed air fighting, long-distance flights and reconnaissance flights are rapidly advancing. Yesterday, for example, I flew to protect our kite balloon from the enemy's airplanes; in the dark, the intensive fire against our airplane conducted by their battery was clearly seen by its lights; it's a very advantageous situation to uncover a well-camouflaged battery, and I naturally took this opportunity, but when you begin to think that these peaceful (as they look from a height of 1,800 meters) flashes can make the airplane all holes, it gives you a terrible feeling, and you are waiting with anxiety for the white and red smoke of their shrapnel explosions.

There were no particularly interesting flights; I took part in one late evening flight in the Voisin. It was very dark for landing and I felt scared when the engine packed up during a sharp turn over a ravine. But for a miracle and the pilot's skill, we would have hit the ravine at full speed, but, thank God, this did not happen. And as you see, I'm writing you a long letter. On one aircraft I put a telescope (comet-finder) received from Backlund; it's very handy in observation, but it is difficult to aim it at the desired object. Since I now fly relatively little I have a lot of time for my studies . . .

Looking forward to hearing from you; I don't need to tell you what a joy it is to receive letters from "there," let alone scientific letters. Give my best regards to the princess; I had my last "cultural" day at your place and I often think about it. Sometimes I wonder whether that cultural day was my last. Then I get scared and don't want to fly, but all this will pass away, within a few years the war will be over.

My exact address: Headquarters of the 9th Army, Aviation Detachment 26, to Ensign Friedmann.

It is pertinent to note that one month after this letter had been sent to Golitsyn the cities and the strongholds of the front mentioned in it were still in the hands of the Russian Army, but as of July 22, they were located almost exactly along the front line. On August 5, Warsaw was abandoned, on the 19th Novogeorgievsk, and soon afterwards all the other cities except Riga (the situation on the northern front stabilized and the Germans failed to reach Riga).

To V. A. Steklov:

July 25, 1915
Army in the Field

My dear and highly esteemed Vladimir Andreyevich,

I haven't written to you for a long time, because our soldiers' life has now become very monotonous and there was nothing to write about; I've now got adjusted, got used to this life, and since I don't expect it to end soon I decided to try and continue my studies, and I do indeed study, although in a rather amateurish way, of course. As you probably know, I'm now an observer-pilot of Aviation Detachment 26 and have to make frequent flights, or rather it used to be so, because now the weather is not very good and there are very few flights. There were two kinds of flight: routine and interesting. The most interesting were the two flights in which I happened to take part in air-fights; the first flight was in a Farman-15 plane, the pilot gave me the controls and for several minutes I steered the aircraft while the pilot was shooting at the German from the carbine; the second air-fight was on a much larger scale: we entered into battle with two enemy airplanes, one of which had a machine-gun; the most fearful thing was to hear the machine-gun firing while it was aimed at you; the distance between the airplanes was extremely small, and I think that it was by a miracle that I escaped death. The fight was going on over our troops' positions and was witnessed by thousands of people; the Germans went away on seeing our second Voisin airplane coming back from a reconnaissance flight. Among the unlucky flights I should mention two flights: during one of them (I was flying in a Farman-16) close to the ground a strong whirlwind threw the airplane, its wing touched the ground and we smashed to pieces – that is, not us but the airplane; the second flight, with a happy ending, but which could have had a sad outcome, was a night flight in the Voisin. Approaching the ground and making a turn near it, we could no longer see the ground because of the darkness; during one of the turns the engine failed, and but for a miracle we would have hit a ravine at full speed; there would of course, have been nothing left of the pilot and me.

I have recently been in Petrograd, went to visit the Pulkovo Observatory and asked Backlund for a small-size comet-finder; he was very kind and gave me an excellent Dollond telescope. It has already been installed in an airplane and the next reconnaissance flight will be done using this telescope: I pin great hopes on it because it has a quite large field of view, magnification and brightness . . .

For the reconnaissance flights I've been awarded the George Cross, but of course whether I get it is a big question. It looks, of course, as if I am really interested in such trifles as awards, but this is the way humans are: they always wish to "play" a little . . . In the detachment, out of boredom, I am learning to fly, but, of course, I will never be a professional pilot: unless the war goes on for ten years or so, but that isn't likely . . .

Don't be angry with me, dear Vladimir Andreyevich, for being so verbose: I wanted very much to discuss the scientific questions from which I am now so far away; I had to stop writing this letter in the middle and fly to defend our kite balloon from the enemy's airplanes. Watching the firing through the binoculars, with every salvo, with every shot of the enemy's battery I thought: "What if the letter remains unfinished, what if in 8 or 10 seconds the airplane is in tatters and falls tumbling down, what if this is the end," but Bernoulli's theorem saved me once again, and we landed more or less safely, and I am writing these lines.

Sometimes I get sick of the war, I want a quick victory, thunderous and overwhelming, but we still have to wait; and yet the spirit is still strong, and if I get used to studying here, then, probably, by the end of the war, I will have finished my dissertation.

Visits to Petrograd

It is interesting to note that Friedmann's autobiography, which we have mentioned many times, was reprinted in 1966 in his *Selected Works* published by the Academy of Sciences, and was based on the text published in 1927. A scrutiny of the original text showed that a number of passages and sentences had been deleted. Among them was this sentence (taken from the section telling about his flights over Przemyśl in December 1914 – March 1915): "Awarded the George Cross for the flights." A. F. Gavrilov writes: "Friedmann ended his rather long front-line service without wounds, but with war awards." Alexander Friedmann's valor appears to have been recognized many times.

Now let V. A. Steklov speak: on September 1 (14), 1914, he wrote in his diary:

Yesterday, at 3 o'clock A. A. Friedmann came. He had come to stay for about three days. In low spirits. I gave him my calculations concerning the deviation of bombs dropped from airplanes and the design of a device for dropping bombs. Friedmann and several other officers and engineers – Lutsenko and Nemchinov (N. N. Andreyev, our student, is also taking part) have set up a small plant producing parts for airplanes. Business is going well, parts have been ordered by the Defense Ministry. He is an industrious gentleman, gets involved with all kinds of projects! He is getting up to six hundred rubles a month, and in addition the plant will pay him several hundred rubles by December. An enterprising fellow!

September 2 (15), 1915. Friedmann had promised to come in the evening, but failed for some reason. This is strange!

November 2 (15), 1915. Friedmann came at 4 o'clock, he has a leave of about two weeks. Says he has got tired. Claims that 500,000 troops are concentrated near Romania and that the army's morale is high. The home front is much worse – a mess. He had lunch with us. He's strange: capable and industrious, but likes to show off, has no clear direction and partly careerist.

November 21 (December 4). At 1 o'clock Friedmann came. He is leaving; had lunch with us. Brought his article asking to submit it to *Comptes Rendus*. All this will be seen to by his wife, and I will only write to Appell. He is leaving on Wednesday.

December 23, 1915 (January 5, 1916). At 3.30 Friedmann turned up, he had been assigned to lecture at a pilot school in Kiev, will stay in the country. Showing off too much! A bad feature.

These last entries clearly show that part of what Friedmann said irritated Steklov. While on the preceding pages Alexander Friedmann was painted solely in rosy colors, here Steklov clearly marked some negative features of his pupil's character. He shared his feelings with Alexander Friedmann's closest friend, Yakov Tamarkin, and did not ask him to keep the conversation secret from Friedmann, rather the opposite. On his next business trip to Petrograd, Friedmann came to see Steklov. This is what Steklov wrote in his diary on February 14 (27), 1916:

At 3 o'clock Friedmann turned up after seeing Tamarkin. The latter had talked with Friedmann about his not always proper behavior; he came to have it out with me. I expressed my displeasure with his growing braggery and boastfulness, with his budding careerism, with his not always gentleman-like attitude to his colleagues (conceit), and his passion for creature comforts, etc. I told him to watch himself and, if possible, to contain himself, to get rid of the base instincts of his (cunning, a tendency to exploit others, etc.). This reprimand will do him good! Hopefully, it will change him a little.

Let us be Friedmann's advocates! A bookworm, who had some, though little, practical experience in the Pavlovsk Observatory, found himself in August 1914 in entirely different circumstances, facing death. Air flights in those years were not safe even in peacetime, and what about wartime?! When they not just *can*, but *want* to shoot you down, to kill you?! The colleagues he had left were doing what was customary for them and him – working at manuscripts and books. And he himself was "working" at the Front near Przemyśl. In the university corridors they met with each other, with their professors, as if there were no war. But he met enemy airplanes in the sky of Galicia, looking the enemy in the face.

And he was also continuing to work, and successfully at that. Out of his reflections and studies, articles are born and published, and in this respect he was by no means lagging behind his friends and colleagues. But apart from this he had the George Cross for valor (maybe not just one), and a quite concrete, visible, tangible job of setting up an airplane workshop, designing new instruments, making up bombing charts so much needed by the Front. A poor college student, then a graduate student hardly making both ends meet, by 1916 he had a salary which was comparable to and sometimes exceeding that of a professor (it was probably at that time, during his visits to Petrograd, that Friedmann collected his superb library with classical German, English, French and Russian works in mathematics and mechanics). And it may well be that he was not boasting to Vladimir Steklov, but was just telling him (with some inner satisfaction and legitimate pride) what he had been, what he became, believing that his teacher would share with him his joy and pride. He addressed requests to his friends because, in the first place, friends are there to help. And, in the second place, one can be sure that he carried out their requests with some readiness and pleasure. In fact, if one is to talk about his later years when he was living in Perm, it can be stated with certainty that it was Friedmann who spared no effort helping Gavrilov and Tamarkin to get there; and again, once in Petrograd it was he, who, having been the first to get into the Naval Academy, brought there Gavrilov, Smirnov and Tamarkin.

Steklov had, in fact, reprimanded Friedmann because he was fond of him, for it is only very close people that really quarrel. Had Steklov been indifferent to Friedmann, and had Friedmann been too conceited and narrow-minded, Steklov would long ago have given him up; but he warned his pupil in a fatherly manner of the dangers of trials by "trumpets" after trials by "fire and air." And Alexander Friedmann could not but admit, no matter whether his teacher's apprehensions were justified or not, that they were motivated by concern about his future. This is why he did not break his relations with Steklov and would continue to seek his advice and support.

Still there is a piece of evidence showing that Alexander Friedmann was a little bit angry with Steklov ("Alexander, you're angry, therefore . . ."). While before February 1916 his letters to his teacher would begin with the salutation "dear and highly esteemed Vladimir Andreyevich," later the epithet "dear" was dropped for a long time and was restored only at the very end of 1918. Later, in 1925, the year of sorrow for Friedmann's relatives and friends, and the year of sadness for science, the way Steklov thought – not about Friedmann's personal traits, but about his impact – was seen from Friedmann's obituaries.

In Kiev

Soon after the retreat in summer, 1915, of the Russian Army on the south-western front, Friedmann, as noted in Steklov's diary, found himself in Kiev. In 1916–17, he gave lectures on aeronautics at the school for observer-pilots which was set up at that time, and in March 1916, he

Alexei N. Krylov.

was appointed head of the Central Aeronautical and Aerological Service of the Front.

Interesting information about this period of Friedmann's activity was found in the file of Academician A. N. Krylov in the Archives of the USSR Academy of Sciences. In 1916, Krylov was appointed Director of the Main Military Meteorological Administration. He went on an inspection mission to some centers of the military meteorological service, and in June 1916, he got to Kiev where he met Friedmann. Krylov made some notes based on this meeting; in addition, he got several important documents from Friedmann. Thus, there was a "classified" list of the members of the Main Military Meteorological Administration drawn up on May 18, 1916. The directorate at that time consisted of 10 people – staff workers of the Main Physical Observatory. Of most interest to us is the assistant director of the department of weather forecasting and local military meteorological stations, M. F. Petelin (P. P. Semyonov-Tien-Shansky headed the department). Krylov's sketchy notes give some idea of the structure of the Russian aeronautical service during World War I. Every army had an aviation unit which was to supply (as well as to check and repair) instruments for six to ten aviation detachments. The detachments, in their turn, conducted systematic weather and wind observations; this was important not only for air flights, but also for the orientation of the headquarters with regard to possible gas attacks by the enemy. The supervision of the entire aeronautical service was the task of the Central Aeronautical Station (CAS) which was based at the time in Kiev and was directly subordinated to the "Most August Supervisor of Aviation and Aeronautics of the Army in the Field, Adjutant-General Grand Duke Alexander Mikhailovich." Ensign A. A. Friedmann was acting head of the CAS. The CAS provided aviation detachments with instruments and manuals on their use, and trained the personnel in the fundamentals of the aeronautical service at a special school which was also located in Kiev and was called "The Military School for Observer-Pilots."

One should note that this was not the only such school in Russia: a similar "Military Air School" existed, for example, in Gatchina; in Petrograd there was the "School of Aeronautics" as well as "Officers' Aviation Courses" at the Polytechnical Institute. A textbook on aeronautics was necessary for regular instruction, and this matter was discussed by Friedmann with Krylov. In the opinion of Ensign Friedmann, which was endorsed by Lieutenant-General of the Navy Krylov, to make up such a textbook – about thirty printed signatures, and at short notice too – was beyond one person's ability. Therefore, a group of authors was brought together ("The Commission on Publishing an Aeronautics Course"), which included mainly staff workers of the Main Physical Observatory. Among them were N. N. Kalitin (in the early 1920s I. V. Kurchatov worked under him in the Observatory), Ye. I. Tikhomirov (the future deputy to Friedmann when the latter was Director of the Observatory), P. I. Molchanov (whom Alexander Friedmann knew by his pre-war work at the Pavlovsk Observatory; Molchanov was the first to design meteorological radio probes), as well as M. F. Petelin. The head of the directorate was to be the editor of the course; Ensigns Kalitin and Friedmann were to be his deputies. The course was to consist of three parts: (1) aeronautical meteorology, (2) aeronavigational instruments, (3) applications of aerology to aircraft control.

In Krylov's records there is a note that a complete course in aeronautics (the publication of which, according to the estimated expenditures, would have cost 10,500 rubles) would take a long time to develop and that it would be very useful to make up an abridged version for the aviation schools. A *Synopsis of Lectures on Aeronautics* was to serve the purpose. There is a reference to this synopsis, which was to be written by Friedmann, in the bibliography of his works, specifying its length (43 pages), and the date and place of publication (1916, Kiev). However, it could not

be found either in Kiev or in Leningrad libraries, including the Library of
the A. I. Voeikov Main Geophysical Observatory, so at times we ques-
tioned whether the synopsis existed at all. But it did! It is attached to the
above-noted file from A. N. Krylov's archival collection, and is, in fact, a
pamphlet. On the cover appears:

Synopsis

of Lectures on Aeronautics
given at the beginning of 1916
at the Military School for Observer-Pilots in Kiev
by Ensign Friedmann
Published by order of the Office of the Most August Supervisor
of Aviation and Aeronautics in the Army in the Field

This is followed by an exquisite decoration with the publishers' imprint
below:

Kiev

Printing House of the Kiev Military District Headquarters
11 Bankovsky Street
1916

The cover has also a handwritten note: "To His Excellency Deeply
Esteemed Alexei Nikolaevich Krylov in happy memory of a visit to the
Central Aeronautical Station on June 11, 1916. The author."

In his foreword to the *Synopsis* Friedmann notes that it was written in
rather unfavorable conditions, when there was no necessary literature at
hand. He also had very limited time. "The only reason for such a hastily
written Synopsis is an urgent need to have within a short time a book
treating aeronautical issues. The author apologizes in advance for the
many slips and shortcomings which the reader will find in the Synopsis,"
and asks in conclusion to address to Kiev, to "the Office of the Supervisor
of Aviation and Aeronautics in the Army in the Field" all comments on
shortcomings and errors found by the readers of the Synopsis, adding that
these comments and remarks will be received with "deepest gratitude."

Since, according to Friedmann, aeronautics is "the science of instruments used for aircraft control and of atmospheric phenomena as applied
to aviation," in his *Synopsis* he provides information about aviation
instruments. It appears that he intended to publish two more parts of the
Synopsis. One of them was to be about aerology, i.e. the theory of
atmospheric phenomena as applied to aviation. This was to include such
questions as vortices in the atmosphere, vertical air flows, air pockets,
turbulence and wind gusts – Friedmann had been concerned with these

questions since his university days. The third part of the *Synopsis* was to deal with the solution of practical aeronautical tasks: methods of orientation, course plotting, choosing suitable meteorological conditions for long reconnaissance flights, etc.[1]

The published first part dealt with the main aeronavigational instruments – altimeter, compass (with a discussion of its deviation and optimal location in the aircraft), tachometer, anemotachometer and clinometer. The *Synopsis* contains very good drawings of all these instruments. It also gives interesting information about the speed of airplanes, hydroplanes and airships of different types including those in which Friedmann himself had flown (Farman, Voisin, etc.). Let us note that according to this information one of the French airplanes had a maximum speed of 160 km/hr.

One may wonder why Friedmann was so well versed in matters concerning various instruments. The answer to this question lies not only in his practical work at the Pavlovsk Observatory, where he was concerned with methods of meteorological measurements (using instruments sent into the atmosphere by kites). There is no doubt that Friedmann had a deep interest in instrument design. As a graduate student at the University he gave much attention to the study of various – though mainly mathematical – devices. Together with V. V. Bulygin and Ya. D. Tamarkin, he studied the instruments at the departments of mathematics (headed by Ya. V. Uspensky) and mechanics (headed by D. K. Bobylev); not content with this they would go to I. V. Meshchersky's department at the Polytechnical Institute.

Alexander Friedmann's departure from Kiev and transfer to Moscow did not allow him to complete his work on the *Synopsis* and write, jointly with his colleagues, a complete course of aeronautics. Yet Friedmann seems to have collected materials for this course. In his library card-catalog there are many titles concerned with the design of various aircraft engines, the fundamentals of aviation, the use of aircraft in wartime, the theory of aimed firing from aircraft, etc.

The Kiev Physico-Mathematical Society

Interesting information about Friedmann's work in Kiev can also be found in the *University News* of Kiev University, mainly in the "Reports

---

[1] In the card catalog of Friedmann's library there is a card with a reference to M. K. Vasnetsov's *Synopsis of Aeronautics* (Kiev, 1916). Could it be the third part of the *Synopsis*? This book could not be found.

and Minutes of the Physico-Mathematical Society of the University for 1915–1916" published in Nos. 9 and 10 of the *University News* for 1917. At that time the Society met five to seven times a year. Its chairman was the well-known researcher in mechanics G. K. Suslov. N. F. Zhukovsky and he were honorary members of the society. Among the prominent scientists who belonged to the society at that time were Ch. T. Bialob-zeski, P. V. Voronets, N. B. Delone and B. N. Delone, D. A. Grave, A. P. Kotelnikov, V. P. Linnik, O. Yu. Schmidt, and others. At the session held on October 19, 1916 and chaired by G. K. Suslov, the following people were nominated for membership in accordance with paragraph 9 of the protocol:

(a) Mikhail Konstantinovich Vasnetsov, an assistant [junior member] of the Astrophysical Observatory of the Emperor's Novorossisk University, teacher at the School for Observer-Pilots;
(b) Alexander Felixovich Gavrilov, working at the Emperor's Petrograd University, an assistant supervisor of calculation of ballistic charts for bomb-targeting;
(c) Lieutenant Mikhail Sergeyevich Gardenin, Director of the Workshop of the Central Aeronautical Station under the Most August Supervisor of Aviation and Aeronautics in the Army in the Field;
(d) Baron Emmanuil Anatolyevich von der Pahlen, Doctor of Philosophy of Göttingen University, Assistant Director of the Central Aerological Station;[2]
(e) Alexander Alexandrovich Friedmann, Master's degree student at the Emperor's Petrograd University in the faculty of pure and applied mathematics, Acting Director of the Central Aeronautical Station.

On October 26, 1916, all these people were admitted to the Society, and Friedmann presented a report "On the organization of a commission at the Aeronautical Station." This commission, according to the minutes of the October 26 session, was elected "to elaborate scientific and practical questions relating to the setting up of a wind tunnel in the estate of St. Vladimir University and to experiments using the tunnel. The members of the commission are Cz. T. Białobrzesky, P. V. Voronets, S. I. Kalyandyk, I. I. Kosonogov, A. P. Kotelnikov, V. K. Rocher, G. K. Suslov, and A. A. Friedmann. The Society decided to grant the commission the right to co-opt new members if the need arises."

B. N. Delone conveyed to the author a few facts about Friedmann's life in Kiev. Friedmann regularly attended the Society's sessions. Boris Delone recalled (in the late 1960s) that Friedmann would usually come in a cab and looked quite impressive in his officer's uniform, with shoulder-straps decorated with a golden eagle (only observer-pilots, who were few

[2] Seven years later, Friedmann was destined to meet again with Pahlen, now in Germany.

in number at the time, had the right to have such shoulder-straps; Friedmann got the rank honoris causa, i.e. without graduating from a specialized school).

In early 1916, as Friedmann himself recalled, he was invited to give lectures at Kiev University and delivered two lectures: "On the motion of a fluid with changing temperature" and "On curvilinear coordinates." He got the title of Assistant Professor of the University, but did not have enough time to give a course of lectures: in April 1917, Friedmann was transferred to Moscow.

6

Moscow–Perm–Petrograd

Moscow

There can be no doubt that interest in Friedmann's personality will increase over the years, and then the many "desert areas" in his biography will disappear, giving place to cultivated "plots" densely populated with events. One such remaining "desert area" is the time that Friedmann spent in Moscow before leaving for Perm. It is known that the Central Aeronavigational Station in April 1917 moved from Kiev to Moscow – together with the personnel and the various services (workshop, instrument store, etc.), and the laboratory. Friedmann was appointed a member of the commission for the construction of an aviation instrument plant. By this time in Moscow several buildings had been prepared for the future plant on the site of former Georgian bath houses. The plant was called "Aviapribor" (aviation instruments); it had started earlier, in 1915, as a workshop for repairing aeronavigational instruments. Friedmann dealt with this very workshop, and he recommended N. N. Andreyev, an acquaintance of his, for a job there. Andreyev started to work there and got in touch with Nikolai Zhukovsky. Later he reminisced that in his first talk with Zhukovsky "Friedmann's name was a key to his heart and he immediately got warmer," and gave an instruction to render Andreyev all possible assistance. Andreyev was interested in wind tunnels necessary for solving the important problem of graduating instruments for continuous automatic recording of wind speed and direction.

In the summer of 1917, Friedmann was appointed head of one of the departments in the plant. The department heads formed a council which operated under the plant's management and collaborated with it in framing the plant's policy, and made technological and organizational decisions. Alexander Friedmann was soon put in charge of the whole plant, becoming its director, thanks to the organizational skill and experience gained during his years at the Front. The plant manufactured all kinds of measuring instruments, as well as devices necessary for high precision bombing and shooting from aircraft (the blueprints were made at the technical section of the plant). The same purpose was served by the charts which were drawn up during 1917 by Friedmann and A. F. Gavrilov, who was head of the computational bureau of the plant. Later the plant ordered a wind tunnel, with a two-meter diameter and a flow speed of around 100 m/sec; Zhukovsky assisted Friedmann and his associates in designing it. According to Andreyev's memoirs, Friedmann came to Petrograd once in a while. In Moscow he and his wife lived with acquaintances in Bolshoi Kozlovsky Lane. On April 10, Alexander Friedmann wrote from there a detailed letter to V. A. Steklov which shows that in Moscow he felt a stranger. Besides, there were unexpected difficulties: right after the October Revolution, when there were great hopes for a long peaceful development of the young republic, the work of the plant was stopped. Alexander Friedmann started to look for a new job and was not sure which of the two options he ought to choose (the Observatory in Petrograd or Perm University). The situation was further complicated by the fact that, as he wrote to Steklov, the war years had told on his health and gave him a serious, though, as he hoped, temporary heart disease. "At the moment," Friedmann wrote, "it does not allow me to move much and travel by train; my doctor says that I cannot yet go to Petrograd." This made it difficult for him to get a job in the Observatory: it required personal negotiations. The time thus freed was devoted to preparation for his future studies in mechanics: he read the famous textbook *Natural Philosophy* by Thomson and Tait ("two T's", as the textbook was called in those distant years), and set himself and solved interesting applied problems, for example, the behavior of a pendulum suspended in a flying aircraft. It was a pleasure for him to write to Steklov that he was helped in solving this problem by the already published Master's dissertation of Ya. D. Tamarkin. In Moscow Friedmann studies this dissertation and the works of English students of mechanics, gets deep into Planck's *Theory of Heat Radiation* – he was seeking there data for constructing a theory of air heating through radiative heat exchange. All these questions later became the subject of Friedmann's studies (which remained partly unpublished). His letter to Steklov seems to end on a sad note: "Forgive me, Vladimir Andreyevich, for writing so much about myself. I'm very depressed; I often bitterly regret taking part in the war; it seems I achieved what I set out to do, but what's the use of it all now?"

Two brief comments on the concluding part of the letter: Friedmann wrote about himself too little; and what he had accomplished became useful very soon, when the Civil War and foreign intervention began, and an air force began to be organized in the Red Army.

Perm

In the pre-revolutionary years the vast territory east of Moscow and Kazan as far as Tomsk contained not a single college or university. This was, naturally, a source of concern for progressively-minded people in the Urals and adjacent areas. Petitions were sent to the Ministry of Education, primarily from residents of Ekaterinburg (now Sverdlovsk) and Perm, with requests for colleges to be opened in those cities, and offers of all kinds of financial assistance on the part of the city authorities, as well as the educated and at the same time propertied part of the population, especially merchants and industrialists.

The report on the opening of a university in Perm, issued by the Perm provincial council (there is a copy in the archives of Petrograd University, in the Leningrad State Historical Archives) is one of the major documents shedding light on this page of university history in Russia. We learn from it that as a result of the many requests from the Urals public, in the autumn of 1915, the second year of the war with Germany and its allies, Professor K. D. Pokrovsky of Yuriev (Tartu) University was sent to Perm. He toured cities of the Urals (Ekaterinburg, Perm, Ufa) to choose a site for Yuriev University in case the situation on the north-western front made it necessary to evacuate it. Such a necessity did not arise, but the project finally went ahead. Officials from the Ministry of Education visited Perm and their reports gave rise to a decision in July 1916 to open a university there. As the first step towards this goal it was decided to set up the Perm Branch of Petrograd University. The Rector of Petrograd University, Professor E. D. Grimm, and other professors, including V. A. Steklov, took an active part in the organizational work of 1916–17 and in later years as well.

The University Branch was opened on October 1 (14), 1916, by Prof. E. D. Grimm. The first Rector of the Perm Branch of Petrograd University was the aforementioned Prof. K. D. Pokrovsky, Doctor of Astronomy. Two hundred and fifty-three greetings were sent to the University; one should think though that not all of them were read out! On October 3, the lectures started. Petrograd University was to provide overall guidance of the Perm Branch, including its Physics and Mathematics Faculty. Its first dean was Prof. B. P. Polenov, who had been a staff professor at

Kazan University; he was a specialist in mineralogy and crystallography. Initially, the Petrograd physicists and mathematicians were represented by people familiar to us from the preceding pages: Prof. A. S. Bezikovich and his wife V. V. Doinikova-Bezikovich, G. G. Weichardt, and others. Forty-nine students signed up for the Physics and Mathematics Faculty.

A. A. Friedmann was elected a professor extraordinary in the Department of Mechanics at the session of the Council of the Faculty of Physics and Mathematics on April 13, which was confirmed by the University Council on May 1, 1918, when Alexander Friedmann had already come to Perm. By that time the University already had on its staff some very young scientists who were soon to become famous: the mathematicians I. M. Vinogradov, N. M. Gunter, R. O. Kuzmin, and the astronomer G. A. Shein.

As in the pre-war years, Friedmann hastened to share his impressions with his teacher V. A. Steklov. On April 27, 1918, he wrote to Steklov that the warm welcome he received at Perm was very much due to Steklov's testimonial which had been sent not long before (on February 19, 1918) to the Faculty of Physics and Mathematics of the University.

Steklov's appraisal of Friedmann's work has been published; we shall only quote its concluding – summarizing – part: "One should note Mr. Friedmann's rare capacity for work, and his general erudition not only in questions of pure and applied mathematics (in the university meaning of the word) but also in many questions of practical mechanics, physics, and meteorology. I have no doubt that Alexander Friedmann, if he devotes himself to scientific work at the University, will extensively develop his scientific activity. Having him as a teacher of mechanics at Perm University is, in my view, highly desirable. The University will find in him a worthy teacher and researcher."

Below are excerpts from Friedmann's correspondence with Steklov. They give a most complete picture of this period of his life (the few publications about Friedmann concerning that period are based on them). The first letter was written in G. G. Weichardt's apartment, where the Friedmanns stayed immediately after their arrival; only two weeks later they rented two rooms in Sibirskaya Street, in the house of a certain Alin: this was a stroke of luck, since there was an acute housing shortage in the city.

April 27, 1918. Perm is surprisingly calm, and everything is done in the city in a family fashion, in a good way, even the training of the Red Army, which is 30–40 strong. After the tumult of Moscow I am having a rest here. And even my diet is much better than in the capital. The university here may be put in very favorable conditions thanks to the vast space available and the complete lack of any possibility of side earnings; the university life seems to be very intensive, although the present lack of premises and instruments is a great hindrance. It's very lucky for the mechanics room that there is a huge gun factory which will assist in equipping and fitting it out. Premises have already been assigned for the mechanics room and there may be other premises assigned at my request on the ground floor, to install heavy machines (for example, for the laboratory of the theory of elasticity and resistance of materials). Of course, one obstacle in Perm will simply be lack of books. We still have to work hard to obtain books, since books are in some sense more important than instruments. I will have to stay in Perm about a month in order to clarify my situation and do something about Alexander Felixovich [Gavrilov – V. F.], I would like very much to get him out of work which is little use.

May 12, 1918. I am extremely happy to have got the job in Perm, because the university atmosphere is very good here – we are involved in scientific work only, and everybody is helpful. The mathematicians are all great friends and very likable people. So I think it's going to be a good place to work, and I will be able to start to pay back my unpayable debt [to Steklov for his favorable recommendation – V. F.] . . . On the plan I have provided, the department will probably arrange the publication of lectures; students here suffer without textbooks. I have other plans lined up, but I deliberately try to contain myself so as not to get swamped, and to have the opportunity to work and pay my debts. A physicomathematical society is to be set up in Perm; we will probably get a big grant and will be able to publish a journal, twenty signatures a year in volume. I will write to you in more detail when the matter gets clearer or I'll tell you when we meet.

June 9, 1918. I have managed, after several days' work, to get the Institute of Mechanics a large two-storied factory building. In this building I intend to set up the main mechanics laboratory for measuring simple mechanical values, then laboratories of material resistance, hydrodynamics, aerodynamics and technical measurement. On the second floor there will be a drawing office, a library with a reading room, a room for charts and reference manuals, collections of models, . . . a classroom and faculty rooms. As you can see, much can be accommodated with some comfort; I want to try and have theoretical mechanics taught in Perm the English way, and I think that it's good when one university differs from another by (so to speak) the teaching style. I know you will probably smile at these intentions, but I don't think you will be angry, because when I speak about a practical direction in mechanics, I am, of course, far from introducing into the university the petty spirit of engineering schools where dabblers are content with collections of formulae riddled with mistakes; I believe that a university should stress the theoretical aspects (in which I include experimental mechanics) illustrated by practical applications of mechanics.

It is seen from the above letters that Friedmann is expecting A. F. Gavrilov to come to Perm. By that time one could say the latter was already a university worker in absentia: using the money donated by

N. V. Mashkov, a rich Perm merchant, Gavrilov bought in Moscow some equipment for the mechanics room (institute) of the University and sent it to Perm, and took part in drawing up the plans for an expedition to study the productive and material potential of this rich area near the Urals. The letters also mention V. I. Smirnov's possible coming to Perm. Friedmann also invites V. A. Steklov himself to come to Perm ("come to see the Perm 'baby' which is our university . . . don't forget your Perm fledglings"), though advises him to wait a bit before coming: there was a real danger of the White Czechs' advancing on Perm.

Let us note one more important confession – a "self-evaluation" contained in the letters. Friedmann writes with humor that some of his "quick grasp of practical matters often renders [him] ill service, because colleagues try to give [him] some responsible practical task;" Friedmann is afraid (and we think not without grounds) that this will distract him from his main calling – scientific research.

Friedmann wished to complete his Master's dissertation as soon as possible, and also to submit another dissertation to the University so as to have the right to give lectures, in accordance with the requirements of that time adopted both at Russian universities and abroad.

In the autumn of 1918 the Physico-Mathematical Society was set up in Perm with its *Journal of the Physico-Mathematical Society of Perm University*. The journal's editorial board included A. S. Bezikovich, G. G. Weichardt, K. D. Pokrovsky and A. A. Friedmann. During Alexander Friedmann's residence in Perm two volumes of the journal were prepared (published in 1919 and 1921). After a long interval its third volume came off the press (1926) with M. Bitov's article "Outline of the activities of the Physico-Mathematical Society of Perm University (1918–1926)." We learn from the article that the first session was held in June 1918. Bitov gives a list of the professors present: K. D. Pokrovsky, A. S. Bezikovich, G. G. Weichardt, A. A. Friedmann, and instructors R. O. Kuzmin,[1] N. S. Koshlyakov,[1] A. F. Gavrilov, G. A. Shein,[2] I. M. Vinogradov,[2] and some others (by 1926 the above persons were no longer working in Perm). The daily affairs of the society were run by the chairman Pokrovsky, the treasurer Bezikovich, and the secretary Friedmann, who had to do a lot of organizational work in publishing the journal, drawing up the Society's charter, and working out the agendas of its sessions. The society received 15,000 rubles financial support from the Narkompros (the People's Commissariat for Education), which ensured the publication of the two

---

[1] Later a Corresponding Member of the USSR Academy of Sciences.
[2] Later a Full Member of the USSR Academy of Sciences.

volumes of its transactions. Local Perm authorities (the Commissariat of Printing) helped hire a skilled compositor and ensure a supply of type: although the articles were saturated with mathematical symbols the typesetting of the journal was of the highest quality! The society was inaugurated on September 24, 1918; it had 60 members and it held its sessions every six or eight weeks. At its sessions, members gave papers on their research; they also dealt with organizational matters. Friedmann presented his paper on the parallelogram of force (published in the first volume) and reported on the state of affairs with the room for textbooks and manuals. In keeping with the traditions of Russian and European universities, members of the society "proposed mathematical problems for solution" – most likely for senior students' contests. Unfortunately, the content of the problems is not known, but we know that they were proposed by Friedmann, Kuzmin and Bezikovich.

In his last months in Perm, after its liberation from Kolchak, Friedmann served as the Rector's assistant for economic affairs;[3] organized the following sections at the University: chemical-technological, civil engineering and electromechanical; and organized two research institutes at the University, in mechanics and geophysics. What a great number of tasks and duties – and among all these jobs undertaken by Friedmann none was a sinecure! It should be added that he was also giving lectures and courses (we shall deal with them later). This was the case with Friedmann before, this was the case with him later as well. But one should remember that taking away time from himself, from his scientific work, he was using it to encourage his colleagues and provide for their research, as well as to train personnel: this was his civic stand which contributed a lot to the formation and development of science in the Soviet Republic. His letter of December 20, 1918 says: "Perm has come under an unlucky star – it has become not just a front-line city, but virtually a city at the Front. There is a rapid, overall and fairly chaotic evacuation. The University is in the second line of evacuation, but no transport or packing materials have been supplied so far, and the evacuation is at a standstill. We are to go to Moscow via Paveletsky Railroad Station, and they say that the University with all its property is to be handed over to Nizhni Novgorod University. I personally am not inclined to leave the city, because all my things and papers are in Ufa where I transferred them from Kazan when there was a rebellion there. If the papers and books were with me I would

---

[3] In the first half of the 1920s in Petrograd similar posts of deputy directors for economic affairs of the Physico-Technical Institute were held by the future Academicians N. N. Semyonov and I. V. Obreimov.

not be afraid of the evacuation, but to leave them means to lose everything; I would miss the works just started most of all . . . It goes without saying that my studies are going very badly under the conditions; this depresses me most of all – the best years pass by and you get nowhere." And in the same letter Friedmann contradicts himself by giving an impressive list of what he did during the short time he was in Perm. There was his intensive work as a lecturer (statics, systems dynamics, differential geometry); giving lectures took ten hours a week, and the preparation for them forty or fifty hours. To the courses mentioned in the letter one should add theoretical mechanics, vector calculation, meteorology, physical geography, a general physics course, and mechanical fundamentals of physics. Besides this, Friedmann conducted classes in meteorology, descriptive geometry and laboratory practice in mechanics and physics. One should also add his own investigations – in dynamics, hydrodynamics, mathematics . . .

Less than a week after this letter had been mailed to Steklov, Perm was occupied by Kolchak's troops.

Studying the Perm archives, K. N. Loskutov managed to find interesting materials dating to the time of the occupation by the White Army under Kolchak. On December 24, Perm was occupied by the White Guard and on January 10, 1919, a session of the University Council was held, at which A. I. Syrtsov, a reactionary professor, blasted the professors, Friedmann in particular, who were sympathetic towards the Soviet power. "In our desire to catch up with the changed conditions after October 25," said Syrtsov, "we accorded special honor to the local bodies of Soviet government – by electing special 'ambassadors,' mediators and interpreters of university affairs, from the members of the Council. This thankless mission was undertaken by Professors Friedmann and [N. P.] Ottokor." And further: "Friedmann wore their doorstep down."

On March 22, 1919 a regular session of the University Council discussed an appeal to foreign universities "On the scourge of Bolshevism experienced by Russia." After a long debate the appeal was adopted on May 15 by seventeen votes to two (with six abstentions). The two who voted against were Friedmann and Weichardt.

In August Perm was liberated by the Red Army. On August 12, Friedmann writes to Steklov:

After Perm was liberated, at first everybody, including myself, fell upon white bread and butter, so all questions of the spirit and less "stomach-related" questions vanished from the life of the Perm intelligentsia. But as the primary animal instincts were satisfied, and as Kolchakovia became more and more Black-

Hundred-minded, on account of the temporary victories of the Whites, there arose a sharp division in the ranks of the Perm intelligentsia, part of which definitely colored itself black, while the other part became more and more suspicious of the "double dyes" of the blossoming power.[4] This division was the source of many clashes in the University Council, but I won't take up your time with accounts of "storms in a teacup" . . .

A series of senseless actions by Perm professors was crowned by the evacuation of Perm University. It fortunately failed and only the staff and a very insignificant part of the cargo managed to reach Tomsk. I remained in Ekaterinburg with part of the cargo and am now back in Perm. To my shame, I must say that I shared the foolish conviction of my colleagues as to the necessity of the University's evacuation; I thought better of it earlier than the others and therefore came back to Perm. The only person who kept his head and saved the remaining property was Bezikovich, who is apparently A. A. Markov's disciple not only in mathematics, but also with regard to resolute, precise and definite actions.

Since Friedmann mentions Bezikovich's name in his letter, it may be pertinent to quote an excerpt from Bezikovich's letter – also, of course, addressed to Steklov, a true and caring father of the Petrograd Pleiad of mathematicians who found themselves in Perm. On September 8, 1919, Bezikovich wrote to Steklov: "During the retreat of the Whites from Perm all the staff of the university were evacuated [to Tomsk – V. F.]. I was the only professor who stayed, along with three assistants: two from the faculty of mathematics, [O. K.] Zhitomirsky and my wife [V. V. Doinikova-Bezikovich – V. F.], and one from the medical faculty. In Ekaterinburg A. A. Friedmann broke away from the Whites."

Alexander Friedmann returned to Perm before its liberation, and Valentina Doinikova remembered how she and her husband stood with Friedmann and his wife on the balcony of the Bezikoviches' apartment, listening to the cannonade and tracking the approaching Red Army by gunfire. According to her, in the last two weeks of Kolchak's occupation of Perm rumors were being spread about the savage reprisals awaiting professors and instructors, and the "atrocities" of Red Army soldiers – Latvians and Chinese.

In his letter Bezikovich, like Friedmann, asks Steklov to help the University to recruit teaching staff. The newcomers will be provided with furnished apartments (even with plates and dishes!). Life in Perm is relatively inexpensive, Bezikovich writes (so Perm residents sent food parcels by people going to Petrograd; specifically, to Steklov and Khvolson). The university equipment was preserved – the situation was bad only with regard to the mathematics library, for which there was hope of

[4] A play on words, the Russian words for "flower" and "color" being the same. The Black Hundreds were right-wing terrorist gangs.

help from Petrograd. Among the advertisements placed in a Perm newspaper (no. 38 of 19 August; its text is quoted in a letter to Bezikovich sent from Perm to Moscow by one of his relations) is this announcement: "The Perm Department of Public Education and the acting Rector of Perm University [i.e. Bezikovich – V. F.] wish to announce that all the property of Perm University has remained in Perm and at present the available staff are making every effort to attract the required scientific staff and to restore the activity of the University to full capacity in the near future." Classes at the University were to resume by September.

As for Friedmann's own letter, he again provides his teacher with an impressive report on what he has done within not quite half a year. One finds here information about his administrative activities, and about his studies in the dynamics of the atmosphere and fluid hydromechanics – Friedmann wants to bring this all together in a dissertation on the motion of a fluid with changing temperature (and wants to present this dissertation in Petrograd). The letter tells of the work of Friedmann's first pupils in Perm: to one of them Friedmann proposed a problem in celestial mechanics, on the movement of a projectile (rocket) between two planets; to another, a mechanical problem on the movement of a material point in a medium with variable density; to a third, on the theory of the altimeter. Friedmann was also busy with the huge job of putting the University back to work.

This aspect of his activity is most fully reflected in the next letter to Steklov of October 6, 1919. Friedmann writes that the first faculties to be reestablished at the University will be those of physics and mathematics, medicine and agriculture, and that Bezikovich and he are particularly concerned about getting staff for the Physics and Mathematics Faculty. The beginning of the classes was shifted to October 15. The junior courses at the Physics and Mathematics Faculty have the full number of students, and enough lecturers and instructors conducting practical classes, theoretical and experimental. "The material conditions," writes Friedmann, "are fairly good here. The salaries are 50 percent less than in Moscow, but the food and fuel are considerably cheaper than in Moscow and Petrograd, so one can get by. The University provides married staff with apartments with lighting and heating (the apartments are very good), and single men with a room on a special floor of the university building where we have now table d'hôte, so single faculty members will be (more or less) provided for as regards meals and will be saved the trouble." Friedmann concludes his letter with sad news about the absurd death of G. G. Weichardt: Ehrenfest's pupil, Friedmann's closest friend in Perm, drowned when swimming in a lake in the summer of 1919.

And the last letter from Perm – of November 19, 1919 – is virtually a cry for help: the University needs teachers! "Especially now we need teachers (professors and instructors) for the following subjects: physics, chemistry, botany, geology, mineralogy and zoology. We beg you to help us – thinking about the spring semester drives us to despair; our former staff are moving away from us at the same speed as Kolchak's front line. It now takes an incredible amount of effort to arrange classes for the first year students; I am to give a general course in physics, and this, as you can easily see, is ruinous both for me and the subject. As for the biological group of subjects, here things could hardly be worse, our only hope is that Petrograd will support us through your assistance. As you can easily imagine, I cannot pursue any of my own studies, because I am fully occupied with university matters. There is no hope that I will have time for my own work this year."

Among the reinforcements the Perm mathematicians got from Petrograd, the most important and happy one for Friedmann was of course Ya. D. Tamarkin, who stayed there about half a year.

7

Theoretical department of the Main Geophysical Observatory

Return to the Observatory

In the early 1920s, Friedmann gets the chance to return to Petrograd. In the winter he gets an invitation to take up a position as a senior research worker – a member of the Atomic Commission at the State Optical Institute – and in the spring he returns to the Main Geophysical Observatory as a senior physicist, and organizes a mathematical bureau there.

Six years before, at the very beginning of the war, Friedmann had been thinking of setting up a theoretical department in the Observatory. In his letters to B. B. Golitsyn we read: "If there is anything that gives me strength now, it's my scientific work and thinking about the Observatory and my future work in it" (May 20, 1915). "We could process computations for this problem using the Ritz method: it will be quite possible to do it in a mathematical department . . . By August I will probably submit to you a reasoned, at least as a first approximation, report on the mathematical department . . . Trying to sort out the details of the organization of a mathematical department I come across various questions of a mathematical nature" (June 25, 1915). He writes about the same to V. A. Steklov: "B. B. Golitsyn promised me, after the war is over, of course, a position as a senior physicist; he wants to put me in charge of the mathematical department which I discussed with you in the spring, and which is likely to be set up at the Observatory. It's hard to think of a better place for me" (July 25, 1915). In 1919, during a short stay in Ekaterinburg, Friedmann came close to implementing this idea at the local observatory. He was elected to the position of "senior physicist in charge of the mathematical bureau," but soon changed his mind and returned to Perm University. And at last, in the spring of 1920, Friedmann's idea and the decision of Golitsyn, the pre-war director of the observatory, were implemented.

Friedmann in the Main Physical Observatory; after 1920.

Friedmann invited young researchers to the theoretical department – I. A. Kibel, N. Ye. Kochin, V. A. Fock, B. I. Izvekov who was slightly older than the others, and L. V. Keller who was 60 by the time he came to work at the Observatory in 1923. "It is to be noted," writes P. Ya. Polubarinova-Kochina, "that in Friedmann's department women were also employed, and he gave them more than 'donkey work'." Besides Pelageya Polubarinova-Kochina, there were V. I. Afanasieva, Ye. M. Zolina and O. A. Kostareva.

Young researchers came to work at Friedmann's department straight after college. Nikolai Yevgrafovich Kochin began to work at the mathematical bureau in 1922, upon his demobilization from the Red Army, while he was still a student at the University, from which he graduated a year later. Ilia Afanasievich Kibel was admitted to the graduate school at the Main Geophysical Observatory in 1925 upon graduation from Saratov University.

How did Friedmann work with his colleagues? His procedure was the one followed today by virtually all research groups, but in those years only by the "youngest" and most advanced ones, like the Physico-Technical Institute (Ioffe's "kindergarten"). All research workers received from Friedmann topics for independent investigation; the scientific seminar was quite active. Kochin was hired as a computation assistant, but he was not given any routine computation tasks (except that he was once involved in a group task of supplementing ballistic charts). Friedmann offered him a topic – the study of a rotating cyclone. Soon Nikolai

Kochin solved two problems formulated by Friedmann in his Master's dissertation *The Hydromechanics of a Compressible Fluid.*

After the interruption caused by the World War and Civil War, foreign scientific journals started to reach Russia. Friedmann and his co-workers were thoroughly studying new investigations by meteorologists and hydrodynamics specialists, mainly of the Norwegian and English schools. Great interest was aroused by the series of works by Vilhelm Bjerknes and his son Jakob, expounding a new theory of cyclone origin and development, which brought back the idea of atmospheric fronts. A detailed review of the theory of cyclogenesis was written by O. A. Kostareva in 1924. Friedmann got interested in the new theory, considering it qualitatively correct, but not sufficiently substantiated mathematically. The seminar participants also reviewed works containing solutions to specific problems related to the frontal model. On March 17, 1924 (the exact date is known from Kochin's abstract published in the first volume of his selected works), Kochin gave a talk on A. Defant's article "On the theory of the polar front." Soon Kochin's article was published, specifying Defant's solution. This article was the start of a major series of works by Kochin, resulting in the creation of the linear theory of cyclogenesis in the early 1930s.

Turbulence was a constant subject of the seminar – from the theory of its emergence in loss of stability, to the theory of intense turbulence. During his review of H. Lorentz's then popular work on the stability of a flow between two parallel planes in relative motion, Friedmann noted that the turbulent flow considered by Lorentz can hardly be dynamically possible. "Friedmann immediately made the relevant calculations, which supported his suggestion," Kibel wrote in his article "On the theoretical determination of the first critical Reynolds number," in which he developed the idea Friedmann came up with at a seminar session.

Polubarinova-Kochina recalls: "Friedmann was one of those administrators who praise their subordinates in other people's presence. If a colleague succeeded in obtaining an interesting result, Friedmann did not stint his praises. Thus, he was in raptures when N. Ye. Kochin discovered a case of adiabatic motion in which vortices are formed, and praising Kochin he added: 'Whereas I was scratching my right ear with my left hand'."

The vertical temperature gradient

The first article published by Friedmann upon his return to Petrograd in *The Newsletter of the Main Physical Observatory* had been completed back in February 1919. It was a development of his paper "On the distribution of air temperature by altitude," written in Pavlovsk in late 1913. Using the standard language of today's scientific evaluations, this series of studies should be recognized as "undoubtedly topical." In 1905, Teisserenc de Bort discovered that at high altitudes the temperature stops falling and an inversion takes place – a change from a negative temperature gradient to a positive one. The stratum of high-altitude inversion was later called the stratosphere, while the transition stratum with a zero temperature gradient was called the substratosphere (tropopause). As B. I. Izvekov wrote in 1926 in his review of Friedmann's geophysical studies, the discovery of the stratosphere "put meteorologists in a great difficulty." The first person to offer a theoretical explanation was K. Schwarzschild – a physicist who later became famous for solving the problem of the gravitational field of a point particle in the general theory of relativity; this solution underlies the concept of gravitational collapses of stellar bodies (black hole formation). In 1906, the general theory of relativity did not yet exist, and Schwarzschild was interested in the physics of terrestrial and stellar atmospheres. He constructed a system of simultaneous differential equations of hydrostatics and radiation transfer. Schwarzschild considered radiation in a so-called "grey" approximation, assuming the coefficient of atmospheric absorption to be independent of the light's wavelength. The optical characteristics of the atmosphere were considered not to vary with altitude. Schwarzschild believed that such a model could explain the high-altitude inversion, but in the papers by W. Humphreys and E. Gold which appeared simultaneously in 1909 it was shown that Schwarzschild's theory did not contain the change of the temperature gradient with altitude, and led to an isothermal atmosphere.

R. Emden succeeded in 1913 in constructing a model describing the high-altitude inversion. In Izvekov's words, Emden considered "not grey, but colored radiation," i.e. took into consideration the dependence of the abosorption coefficient on the light's wavelength. According to him, radiation had "two colors:" the absorption coefficient assumed two values, one for short and the other for long waves. With this approximation, and accepting, as Schwarzschild did, the hypothesis of radiation balance, Emden solved the problem and obtained temperature gradients close to the ones known from aerological observations in the stratosphere. However, Emden's solution resulted in two large gradient values close to the ground surface. "This gave A. A. [Friedmann] an idea," Izvekov writes in the above-quoted article, "that true equilibrium in the lower strata is impossible and is disturbed by vertical flows." Friedmann included stationary vertical flows in his model and in order, firstly, to study the effect of this mechanism "in pure form," and, secondly, to reduce the attendant mathematical difficulties, he returned to Schwarz-schild's "grey" atmosphere. The problem was reduced to a complex differential equation, which Friedmann integrated using expansion into a power series of a small parameter proportional to the vertical velocity. It turned out that convection too brings about the phenomenon of high-altitude inversion, computed temperature gradients being in good agreement with those observed near the surface of the earth, though less in agreement with experimental data on temperature dynamics in the upper layers of the atmosphere. Concluding his 1913 paper, Friedmann wrote: "The partial explanation of air temperature distribution by altitude presented here is, of course, far from the truth; to approach this goal one first needs, a larger body of experimental material, and one should then take into account the dependence of the absorption coefficient both on the wavelength and on the altitude of the absorbing stratum." Friedmann the physicist saw the shortcomings of his model, and the ways to eliminate them, while Friedmann the mathematician was concerned about the incompleteness of the solution. In the German abstract of his paper – the article in *Geofizichesky Sbornik* (The Geophysical Collection) was written in French – he notes: "I did not manage, unfortunately, to prove the convergence of the power series in μ. The well-known Poincaré theorem cannot be applied in this case." Friedmann's teacher, Academician Steklov, considered this shortcoming quite excusable. In his review of Friedmann's works submitted to the Faculty of Physics and Mathematics of Perm University in February 1918, when Friedmann was standing for a professor's position, Steklov, noting Friedmann's priority in taking the vertical flows into consideration, also mentioned the deficiencies of the work: "Although from the theoretical point of view the solution does not have all the rigor of precise analysis, in questions of this kind it is impossible to demand absolute rigor." The "indulgence" issued by Steklov did not calm the "Master's student of pure and applied mathematics." In the autumn of 1915, busy with preparing his lecture courses, organizing the Physico-Mathematical Society and publishing its journal, Friedmann finds time to continue his scientific work. In September 1918, he informs Steklov that he has found a system of functions with which it is convenient to expand the solution into a series, and has managed to prove the convergence of the series. In his December letter he again mentions this result and says that he has prepared an article for publication. The written article was waiting to be published, while the author went through the retreat of the Red Army, saw "the double dyes of the blossoming Kolchakovia," "shared the foolish conviction of his colleagues" concerning the evacuation of the University during the White Army retreat, "thought better of it earlier than the others" and got back to Perm. During all these tribulations all Friedmann's manuscripts were lost, as well as most of his library. This may be the reason why on coming back to Petrograd he published only a short pre-publication announcement promising to publish a detailed article in *Geofizichesky Sbornik*. But the article never came out, and in later years Friedmann neither returned to the subject himself nor proposed it to any of his pupils.

Why did Alexander Friedmann lose interest in this problem, which had occupied his thought for many years? It is difficult to give a definite answer to this question. First of all because, if we may say so, the theoretical and experimental status of this study are not altogether clear. Friedmann's articles do not indicate to what space and time scales this solution can be applied. If he meant an area of small horizontal extension at short periods of time, then the neglect of horizontal components of velocity, which is required to obtain the solution, makes the model too rough, so that its further elaboration, after the possible effect of vertical flows had been demonstrated, is indeed superfluous. At present, this problem is interpreted in a different way, namely as a problem about the mean distribution of temperature over a long period of time. With such an approach Friedmann's articles become the first Russian investigations in "theoretical calculation of the mean climatic values of meteorological elements,"[1] but at the same time they become vulnerable to criticism. In publications dedicated to his centenary this criticism was oblique: "Friedmann's ideas were, as a rule, indirectly taken into account by considering the turbulent heat transfer, i.e. the transfer of heat by unordered air flows. But the present state of climate theory and computer technology allows us to put on the agenda the task of calculating climatic values of meteorological elements by integrating the equations of hydromechanics, moisture transfer and radiation theory, taking into account *all* major factors including convective heat transfer by ordered vertical flows." The above-quoted article by Izvekov, indeed, shows that Friedmann and his pupils had all the necessary information for "introducing turbulent heat transfer into the analysis:" they knew about the stochastic nature of vertical flows connected by vortices with the horizontal axis, and they were aware of the

[1] M. I. Yudin, "Alexander Alexandrovich Friedmann," in *Izvestia Akademii Nauk SSSR* (Bulletin of the USSR Academy of Sciences), Geophysical Series, No. 7, 1963, pp. 1087–99.

model of vortex (turbulent) heat transfer, developed by Taylor and used by him to explain some temperature inversions. Finally, speaking about 1925, among the department's staff workers was Kibel, who solved this problem in 1943. One "very little" thing remained to be done – to change the perspective. One might paraphrase this passage from Einstein's *Autobiographical Notes*: "Why were another seven years required for the construction of the general theory of relativity? The main reason lies in the fact that it is not so easy to free oneself from the idea that coordinates must have an immediate metrical meaning." And then why did it take another eighteen years to develop a correct theory of the vertical temperature gradient? It was not so easy to understand what specific problems one was to solve. As soon as this was understood, Kibel got a very simple solution which was in fairly good agreement with experimental results for all latitudes, and in good agreement for latitude 40°. Refinements were still to be made. In 1947, Ye. N. Blinova introduced horizontal transfer into the model and obtained agreement for all latitudes. (Here another story begins – together with Blinova's article the journal published L. R. Rakipova's article which discussed the effect of space dust on the whole phenomenon. Forty years have passed, and now the problem of temperature distribution and the effect on it of solid particles – no longer of space dust, but of soot from fires which may be caused by nuclear explosions – is the central problem in studies of "nuclear winter.")

The analogy between the theory of relativity and the problems of the vertical temperature gradient may seem unjustified, yet "the search in the dark with its tense suspense" (Einstein) is involved in solving both basic and particular problems in the sciences. Besides, this comparison is justified by the fact that two years later it was Friedmann who managed to "get rid" of notions which were hindering progress and were shared by Einstein himself.

"The Hydrodynamics of a Compressible Fluid"

While the series of studies on the theory of the vertical temperature gradient was interrupted, the other series begun during the World War with the article "On turbulence in fluids with changing temperature" was further developed. In the autumn of 1920 and the winter of 1920–21 Alexander Friedmann was working on his dissertation *The Hydrodynamics of a Compressible Fluid*. Professor I. V. Meshchersky of the Polytechnical Institute, in his paper presented at a joint meeting of the physics section of the Russian Physico-Chemical Society and the Physico-

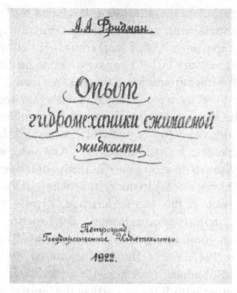

Title page of the lithographed edition of *The Hydromechanics of a Compressible Fluid.*

Mathematical Society, called this dissertation Friedmann's major work in hydromechanics. Friedmann's dissertation, and related studies by his pupils, considered a broad range of theoretical problems originating in "the extensive area of atmospheric motions."[2] The study of fundamental problems in the hydrodynamics of a compressible fluid was geared to the needs of dynamic meteorology.

Friedmann stressed that he was studying the generalized motion of a compressible fluid, understanding by this the rejection of the barotropic hypothesis – the dependence of pressure only on density. Barotropic flows, although leading to more complex problems than incompressible fluid flows, are in many ways similar to the latter. First of all, Helmholtz's two theorems – on the persistence of vortex lines and intensity of vortex tubes in the fields of conservative forces – hold for flows of both types. For baroclinal flows, i.e. flows in which the surfaces of equal pressure (isobars) and lines of equal specific volume (isoteres) do not coincide, but intersect, Helmholtz's theorems do not hold; the problem of "vortex tube destruction and patterns of tension change in these tubes" becomes the main one. Barotropicity is a consequence of isentropicity, i.e. stability of entropy across the whole flow. This condition is violated if there is an

---

[2] This and other quotations in this section are taken from Friedmann's dissertation.

inflow (outflow) of energy, which takes place in the atmosphere, although the deviation from the adiabatic condition is not very great. Friedmann emphasized that adiabaticity is not equivalent to isentropicity (different particles may have different entropy). In any case, the observations accumulated by that time pointed to the significance of the baroclinal factor in atmospheric dynamics, and demanded close attention to the problem of generation and destruction of vortices.

For convenience' sake, in his analysis Friedmann divided the problem into two parts, kinematic and dynamic; thus the dissertation had two parts, "The kinematics of vortices" and "The dynamics of a compressible fluid."

A wide use of vector analysis was a significant feature of Friedmann's treatment of kinematics. From the reminiscences of Friedmann's pupils and associates at the Polytechnical Institute, L. G. Loitsyansky and A. I. Lurie, we know that he was a pioneer of the vector-based interpretation of solid-body kinematics. Vector analysis was all the more useful in fluid kinematics. "The language of vector analysis not only essentially simplifies calculations and formulations of theorems, but makes it possible to carry out the study in a symmetrical fashion, concentrating not on the accidental features of the chosen coordinate system but on the essence of the matter – the motion of the fluid." Friedmann introduced some new differential vector operations – the Helmholtzian and full Helmholtzian (the full Helmholtzian of a vortex is the right-hand side of the Helmholtz equation in the general case, and the Helmholtzian is that in the case of an incompressible fluid). The Helmholtzian **H** and the convective vector **D**, which is the difference between the total derivative of a vortex and its Helmholtzian, play an important role in Friedmann's kinematics of vortices – it is through them that differential and geometrical character-istics of vortex changes are expressed. To introduce the latter, Friedmann considers the following three directions at two successive instants of time: the direction of the vortex line at the first and second instants (directions of the "old" and the "new" vortices) and the direction of the fluid line to which the initial vortex line has been transferred by the second instant. These three directions form a trihedron which, under the conditions of Helmholtz's first theorem, is transformed into a plane angle. The ratios of the plane angles of this trihedron to the time interval Δt between the first and second moments have finite limits which characterize the vortex change. Vortex deviation is the limit of the ratio of the angle between the new vortex and the fluid line to Δt, vortex bend is the limit of the ratio of the angle between the old vortex and the fluid line to Δt, vortex inclination is the limit of the ratio of the angle between the old and the new vortices to Δt, vortex torsion is the limit of a bihedral angle of the circumscribed trihedron with a rib coinciding with the fluid line. Calculation of these characteristics using vectors **H** and **D** is the subject of the second chapter of the first part, in which, according to Meshchersky, vortex change is "exhaustively" studied.

The second part of the dissertation also consists of two chapters. The better known of the two is the chapter "Conditions for the dynamic possibility of motion," meaning the equations of hydrodynamics without "dynamic elements," i.e. pressure and density. This exclusion has the following practical meaning. Solving direct problems of hydrodynamics – determining velocity, pressure, density and temperature with given volume forces, heat sources, initial and boundary conditions – presents great difficulties; it became possible only with the advent of powerful computers. But in many cases it is possible to guess the kinematics of the flow – the dependence of velocity on the coordinates and time, which contains a certain number of arbitrary functions and constants. The question arises whether such a kinematic form is possible, i.e. can one determine the dynamic elements with such a distribution of velocities? The conditions under which the hydrodynamic equations can be solved with regard to pressure and density are, in fact, "conditions for the dynamic possibility of motion." For an incompressible or barotropic fluid these conditions are provided by the Helmholtz equation, which results from the equations of motion with pressure excluded. In the baroclinal case one should exclude not only pressure but also density and temperature; moreover it is necessary to take into consideration not only the equations of motion and continuity but also the energy equation.

In his dissertation, having noted the need for such an approach, Friedmann confined himself to a less general task – the exclusion of pressure and density from the equations of motion and continuity. The conditions thus obtained are necessary, but not sufficient, for the possibility of motion; the obtained distributions of kinematic and dynamic elements should also satisfy the energy equation. Later, Kibel considered the generalized form of Friedmann's problem, taking into account the equation of heat influx, and obtained conditions which he called thermodynamic conditions for the possibility of motion (Kibel's article was published in the third issue of volume 5 of *Geofizichesky Sbornik* in 1932).

In the above-mentioned article "On vortices in a fluid with changing temperature" published in 1916, Friedmann obtained conditions for the dynamic possibility of motion in the form of nine scalar equations. He showed in his dissertation that only five of them are independent. Depending upon the way the fields of pressure and density are restored by the given velocity field, Friedmann classifies the motion of fluids (in Kibel's paper this classification is slightly simplified). In section 16 of his *Hydromechanics* Friedmann considered a number of cases of conditions for the dynamic possibility of motion. Most of these examples were related to the construction of the hydrodynamic model of a cyclone. From 1917 onwards, a number of papers on this subject were published by English meteorologists and students of hydrodynamics. While in the older studies (Overbeck, Marquis) the problem had been solved by assuming an incompressible fluid, and stationary models had been constructed, the "new wave" was already constructing a moving cyclone in a compressible fluid. However, all English researchers resorted to various simplifying assumptions. Lord Rayleigh neglected the vertical component of the Coriolis acceleration; N. Shaw considered geostrophic flow (isobars coinciding with flow lines); Green confined this task to the case of isothermal flow. Friedmann set the task of getting rid of these limitations. The approach he developed made this possible, but, as it turned out, the chosen class of arbitrary functions by which the cyclone model was described was not wide enough; in this class a moving cyclone turned out to be possible only for an incompressible fluid. In his memoir "The theory of compressible fluid motion" which was completed in September 1922, though not published in *Geofizichesky Sbornik* until 1927, after his death, Friedmann suggested a method by which a cyclone could be "set in motion" in a compressible fluid: "the reason for this limitation appears to be that we have assumed ζ [angular velocity – E. T.] to be constant, independent of altitude and time. Assuming that ζ is a function of the altitude z, we will see that there are baroclinal motions of a rotating fluid which are dynamically possible. In this case cyclones and anticyclones can be constructed, with the center moving in some arbitrary fashion, with curved isobars and, of course, with variable specific volume depending on the altitude." Friedmann writes in a footnote that "the case of rotational flow where ζ depends on z and t was studied in detail by N. Kochin, a staff worker at the Main Physical Observatory in Petrograd, who obtained baroclinal motion (cyclone and anticyclone with mobile center)." It may be that in giving the task to his young colleague the professor already "knew the answer" and pretended later that he was distressed about his own "lack of resourcefulness." Polubarinova-Kochina's reminiscences allow this possibility: "Self-criticism was, in fact, characteristic of Friedmann, and he often expressed regret over his shortcomings, real or imaginary."

However that may be, in Kochin's work (published in 1924) the series of investigations by Rayleigh, Shaw and Friedmann was completed – a model of a cyclone (to be more exact, of a cyclone nucleus, because in Kochin's solution the velocities increase infinitely when receding from the axis and he has to "cut" the solution artificially at a certain distance from the cyclone center) was constructed. The validity of the model was supported by the qualitative agreement of calculated trajectories of air particles with the actual trajectories observed by Shaw and Lempfert – both had characteristic loops and points of return.

In September 1924, Nature set up a grandiose experiment which showed that the problem of cyclones is extremely "topical" and, unfortunately, rather far from being solved. A flood occurred in Leningrad, one of the largest in the history of the city. It had not been predicted and was totally unexpected for both the authorities and residents of the city. The public was expressing its anger, and the Acting Director of the Main Geophysical Observatory, Ye. I. Tikhomirov, "was extremely busy, having to take part in many meetings, sitting on commissions and investigations connected with the calamity which had afflicted Leningrad." Friedmann was painstakingly thinking whether meteorology was or would ever be able to help people foresee such a catastrophe. He sent Tikhomirov a letter at the time, presenting his thoughts and doubts. Seven years later Tikhomirov, sorting out his papers, found the letter, decided that "it had not lost its significance," and had it published in the *Meteorological Bulletin* under the title "Can microcyclones be predicted?"

I remember discussing with you [wrote Friedmann] one difficulty which exists in present-day dynamic and synoptic meteorology, and even threatens its future development. Very many meteorological phenomena causing great damage are determined by vortex systems whose diameter (not of vortices but of the system) has an extremely small size (around a hundred kilometers). These are the vortex systems which are close to the rotating fluid model of Rayleigh and Shaw, with which N. Ye. Kochin and I have concerned ourselves so much.

The difficulties these systems present for dynamic meteorology lie first of all in the extremely capricious character of their motion and in the extreme sensitivity of the patterns of change with time of meteorological elements in these sytems. Their motions resemble, to a certain extent, the capricious trajectories of atmospheric particles described by Lempfert and Shaw for trajectories of vortices moving under the influence of various vortex systems. Today it is absolutely impossible to predict the movements of these small-scale vortex systems, owing to the lack of a sufficiently dense network of stations continuously monitoring meteorological elements, as well as characteristics of turbulence, not only on the ground, but also in the free atmosphere.

Let us assume, though, that in future we shall have a sufficiently dense network of stations fully capable of making the above-mentioned observations. The question arises if it will be possible under these ideal conditions to foresee the movements and changes of the above-mentioned fierce but small-scale vortex systems. The answer to this question may depend entirely on how far ahead we would like to foresee the phenomenon. Of course, in the case of a very small time period, under the indicated ideal and almost unattainable conditions, prediction is possible. But it becomes impossible once the time period increases, yet it is only this latter case that is of significance in practice.

Let us interrupt reading the letter, which combined thoughts about the most complex problems of hydrodynamics, plans of the future Director of the Observatory, and a feeling of frustration because of inability to produce the exact weather forecast which was so necessary. Let us go back to Friedmann's dissertation.

The last chapter of the *Hydromechanics* was devoted to the study of the relation between kinematic and dynamic elements. From the practical point of view it would be extremely useful to express velocity components in terms of pressure, density, temperature and their derivatives, since dynamic elements are much easier to determine from measurements than are velocity components. In such an "extremist" version the problem proved to be insoluble. Friedmann found the following set of elements in terms of which the velocity of the flow is determined: dynamic elements of the first order, pressure and density; dynamic elements of the second order, intensity of energy influx and velocity divergence; and, finally, the scalar product of the so-called turbulizing vector (which is the vector product of pressure gradient and density) and the vortex vector. He failed to express the latter element in terms of dynamic ones; the question whether this is impossible, or the required conversion of equations was simply not found, was left open by Friedmann.

Even after the completion and successful presentation of his dissertation, Alexander Friedmann did not lose interest in questions on the dynamic possibilities of motion. An interesting synthesis of the idea of "dynamic possibility" with the idea of simplifying the equations of hydromechanics was provided in the lecture "On approximate conditions of dynamic possibility of motion" given by Friedmann at a seminar of the theoretical department. Assessing the orders of quantities comprising the first condition of dynamic possibility (i.e. that the turbulizing vector equals the vector product of the dynamic gradient and the gradient of the specific volume logarithm) Friedmann showed that this vector equation can be replaced by its projection onto the vertical axis. It is interesting to note that the equation obtained by Friedmann is equivalent to the second approximation in the asymptotic expansion of equations of hydrodynamics in terms of the small parameter known as the Rossby number (the root of the ratio of kinetic and potential energy of the atmosphere). The first approximation in this asymptotic process is not closed, since the continuity equation turns out to be a consequence of the geostrophic correlation emerging in this approximation. A closed system of equations (so-called filtered equations, since they do not contain rapid processes of wave propagation) is provided by the second approximation. That this system was obtained by Friedmann is not surprising: the order assessments he used and the fact that the Rossby number is small reflect one and the same "atmospheric reality." Friedmann's paper "On approximate conditions" appeared in *Geofizichesky Sbornik* after his death. It was edited and prepared for publication by Kochin. This was one of the first steps Alexander Friedmann's pupils took to preserve and develop his legacy. The extension of Friedmann's investigations by Kibel to the case of motion with a given influx has already been mentioned. This work was preceded by Izvekov's article, published in 1924, in which the conditions of dynamic possibility of motion were obtained for a viscous fluid. In 1934, Izvekov, Kibel and Kochin prepared a printed edition of the *Hydrodynamics* (in 1922, a small number of copies had been published in a lithographic form). The supplement included in the book presented all the results developing and summarizing Alexander Friedmann's ideas.

After the 1934 publication Soviet researchers in hydromechanics seem to have decided that this field had been exhausted, and papers on the dynamic possibility of motion no longer came out. In the 1950s, non-Soviet researchers, not aware of the work of Friedmann's school, attacked this problem. R. Berker considered again the problem of excluding pressure, density and entropy in the case of adiabatic flow, and G. J. Eriksen obtained interesting results for a particular case of a steady plane flow of compressible gas. Friedmann's equation describing a change of vortices was obtained anew by A. Vajonyi (an insignificant difference was that the equation was given not in Helmholtz's form but Beltrami's). In his well-known work *Mathematical Principles of Classical Fluid Mechanics* (see bibliography), J. Serrin cited papers by Vajonyi, Berker and Eriksen, being unaware of the series of studies by the Leningrad school.

In the Soviet Union *The Hydromechanics of a Compressible Fluid* is generally recognized as a classic study, but its content began to be forgotten. In the collective monograph *The Development of Mechanics in the USSR, 1917–1967*, published in the series *Fifty Years of Soviet Science and Technology*, Friedmann's dissertation is mentioned twice – in the 11th chapter, "Dynamics of an ideal fluid," and in the 13th, "Gas dynamics." In the former a detailed account of the dissertation is given, although without mentioning the fact that Friedmann generalized Helmholtz's results to include non-barotropic flows. As for the 13th chapter, it even contains a statement that Friedmann's work is devoted not to a compressible, but only to an inhomogeneous fluid: "Friedmann introduced the notion of, and derived conditions for, the dynamic possibility of motion of a compressible fluid in the general sense (density is an unknown function of coordinate and time)." It is unclear how this misunderstanding (perhaps a result of the above-quoted book) could gain such wide acceptance. Historians of science have yet to determine the true value of the work of Friedmann and his school in the development of hydrodynamics and dynamic meteorology.

The theory of turbulence

The scientific destiny of the other area developed in the theoretical department of the MGO – the correlation theory of turbulence – turned out to be more fortunate. It was presented in the paper by A. A. Friedmann and L. V. Keller at the congress in Delft in 1924 (a more detailed account was published by Keller in *Zhurnal Geofiziki i Meteorologii* in 1925).

The invitation of the 60-year-old Keller, who had no name in science, to work at the Observatory is without precedent in the history of scientific institutions. The life of Lev Vasilievich Keller was far from being simple. He graduated from the Faculty of Physics and Mathematics of St. Petersburg University in 1884 with the degree of Candidate of Mathematical Science. As an undergraduate, Keller wrote a paper "On the distribution of matter in interplanetary space" and was given an award for it by the University Council. But soon after his graduation his scientific work was interrupted. Keller had not served a year as custodian of the practical mechanics room in the faculty headed by Prof. D. K. Bobylev, when he was dismissed from the University by the Minister of Public Education (Delyanov) for taking part in a students' meeting. Keller went back to his native place in the Crimea and got involved in vine growing. In 1893–94 he attended lectures at Berlin University. In 1893 he began his many years of service as a statistician at local and government institutions. Dealing with insurance problems, Keller mastered the methods of mathematical statistics. This was the reason why he was invited to work in the theoretical department of the Observatory. Friedmann clearly saw that statistical methods were needed to study atmospheric turbulence. He came to believe in the elderly statistician unknown to science. The years to come showed the brilliant success of Alexander Friedmann's "cadre policy."

Treating fluctuations of hydrodynamic values in a turbulent flow as random functions, Keller and Friedmann described them by introducing correlation functions between the values of one and the same or two different elements at different points of space and/or at different moments of time. Singular-point correlation functions coincide with additional turbulent tensions introduced by O. Reynolds. The system of averaged hydrodynamic equations containing Reynolds' tensions turns out not to be closed. Keller and Friedmann attempted to close the system by adding it to new equations describing the change in time and space of Reynolds' tensions themselves. The new equations in their turn contain a number of additional unknown quantities and again do not form a closed system. To close the system, Keller and Friedmann suggested a hypothesis which holds for Gauss's random fields. As a result, they obtained a closed system of equations with respect to twenty unknown functions. However, the most significant result Keller and Friedmann obtained was not the system itself, but the method they used to obtain it. Generally speaking, using this method one can construct an infinite system of equations with respect to the highest moments of the hydrodynamic variables – the Keller–Friedmann chain. If one breaks this chain at some finite step, one will obtain a non-closed system of equations which nevertheless imposes certain dynamic relations upon statistical characteristics of turbulence. Various hypotheses (for Reynolds' equations chiefly the hypotheses of Karman and Prandtl, and for the more general kinds those of Keller and Friedmann) based on plausible reasoning and experimental determination of some constants, delineate the model of turbulence. As was found later, a closed equation can be constructed, not with respect to usual functions but in variational derivatives with respect to characteristic functionals. This equation, equivalent to Keller and Friedmann's infinite chain, was obtained by Eberhard Hopf in 1952.

Unlike Friedmann's studies in "deterministic" hydrodynamics, the ones in statistical hydrodynamics had almost no further development in his school. Only L. G. Loitsyansky, who continued Friedmann's work of developing hydrodynamic specialization at the Polytechnic Institute and set up there a faculty of hydroaerodynamics, obtained in 1939 an important result in the correlation theory of homogeneous isotropic turbulence (this notion was introduced by Taylor in 1935). Loitsyansky found that under certain conditions in a homogeneous isotropic turbulent flow a certain value is conserved which is proportional to the square of the full moment of fluid momentum ("Loitsyansky's invariant"). In the early 1940s, the initiative in the field of statistical hydromechanics in the Soviet Union passed to the Moscow school, where A. N. Kolmogorov had developed the theory of the universal stationary mode of small-scale components of turbulence at large Reynolds numbers (close results under special constraints were obtained by A. M. Obukhov). Trying to solve the system of equations obtained by Keller and Friedmann was out of the question in the mid 1920s. This is an unsolvable task even for the present-day level of computer technology. Even a more modest task – verifying the hypotheses used in deriving the system, based on the statistical material related to macroscopic turbulence – was beyond the capacity of the small staff of the theoretical department. But the incorporation of the basic result obtained by Keller and Friedmann into theoretical hydrodynamics was a foregone conclusion – it was an important step in the right direction.

Professor Friedmann and his group of specialists in theoretical hydromechanics elaborated a wide range of questions, both particular and general, some promising and some leading to a deadlock. Most of them proved to be general and promising. The work accomplished at the theoretical department during Friedmann's lifetime would alone give Friedmann's school a prominent place in the history of science. Yet, in the years to come there were to be Kochin's theory of cyclogenesis, Kibel's method of short-term weather forecasting, Polubarinova-Kochina's work in the theory of filtration, and the work of their pupils. One can only speculate what Friedmann's school would have accomplished, but for the untimely death of its founder.

8

Space and time

With Friedmann's return to Petrograd, the last and exceptionally eventful and productive period of his scientific activity began. His talent reached maturity, gained confidence in itself and full strength; yet his desire for knowledge, his hunger for creative effort seemed to be insatiable. Friedmann would tell his relatives and friends: "No, I'm an ignoramus, I know nothing, I should sleep less and should not do anything outside science, because all this so-called 'life' is a mere waste of time." He was on the threshold of the most important accomplishment in his life – he was soon to develop his cosmological theory.

Friedmann's long-time and close friend, the mathematician (and later Professor) Alexander Felixovich Gavrilov, wrote: "The last five years in the life of this amazing person were full of virtually selfless work in new areas, with tremendous achievements."

Studying relativity

Friedmann set out to study the general theory of relativity with unusual diligence and at the same time with great interest and zeal. At that time this theory was referred to as the strong relativity principle. That was by no means his only passion at that time, but the theory of relativity undoubtedly overwhelmed him with its broad scope, its simple and clear theoretical basis, its elegant mathematical apparatus. The structure of the Universe as a whole had for the first time become the object of exact scientific study. The nature of space and time was linked in the new theory with the distribution and motion of gravitating masses in the Universe.

The general theory of relativity was developed by Albert Einstein in 1914–16. Cut off from world scientific literature by the years of war and revolutions, and the blockade of Soviet Russia, Soviet scientists got

Vsevelod K. Frederiks.

acquainted with the new theory only after several years' delay. And Friedmann was the first to write a study on general relativity, which was published in 1922. A year earlier the first scientific survey in Russian of the bases of the general theory of relativity had been published in the review journal *Uspekhi Fizicheskikh Nauk* (Advances in Physical Sciences). It was written by Friedmann's colleague, a professor at Petrograd University, Vsevolod Konstantinovich Frederiks, and was based on papers and lectures delivered by Frederiks in Petrograd and Moscow to acquaint Russian physicists with the new theory. In 1923, a professor at the Polytechnical Institute, Yakov Ilyich Frenkel (later a Corresponding Member of the USSR Academy of Sciences) published a book *Theory of Relativity* (Mysl Publishers, Petrograd, 300 pp.). Its foreword, dated November 1922, says: "The present book is the first (in the Russian language) 'non-popular' handbook on the theory of relativity . . . It was written almost a year ago and formed the basis of a course of lectures I delivered in the summer semester of this year to students of the First Polytechnical Institute."

From the 1920s there was a regular seminar at the Physical Institute of Petrograd University, where Friedmann and Frederiks presented papers on general relativity. According to Vladimir Alexandrovich Fock, a participant in the seminar and later a Full Member of the Academy of Sciences, the styles of their presentations were different. Frederiks put the main emphasis on the physical side of the theory, but did not like mathematical formulations, and tried to make his presentations physically substantive and more qualitative. Friedmann, on the contrary laid emphasis not on physics, but on mathematics, striving for rigor, exactness and completeness in formulating, elaborating and discussing problems. There were frequent discussions between Friedmann and Frederiks, which were particularly interesting and instructive for the young participants in the seminar, including senior university students. Professors Yuri Alexandrovich Krutkov and Viktor Robertovich Bursian also took part in the seminar and sometimes gave presentations. Among the attendees were colleagues of Friedmann, mathematicians and specialists in mechanics who worked with him at the Polytechnical Institute or

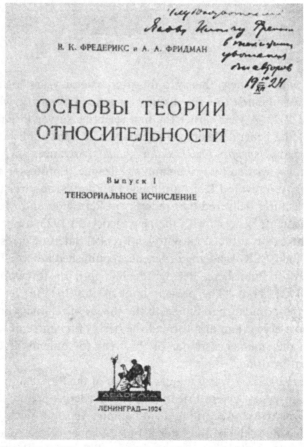

Title page of *Fundamentals of the Theory of Relativity* by V. K. Frederiks and A. A. Friedmann.

the Main Geophysical Observatory. A characteristic phrase at that time was: "We will soon understand these questions – Friedmann has set out to master the theory of relativity!"

At about the same time, Friedmann and Frederiks began to write a fundamental monograph on the theory of relativity. They set themselves the task of presenting the theory with adequate rigor from the logical point of view, assuming the reader's background in mathematics and theoretical physics did not exceed the level of knowledge given by Russian universities and higher technical educational institutions. It was originally intended to publish the whole book at once, but technical obstacles made the authors divide the book into five parts and prepare these parts as separate issues. The first issue of the book expounded the fundamentals of tensor calculus. The second issue was to be devoted to the fundamentals of multi-dimensional geometry, the third to electrodynamics, and, finally, the fourth and fifth to the fundamentals of special and general relativity.

Unfortunately, only the first issue of the whole monograph came out: V. K. Frederiks and A. A. Friedmann, *Fundamentals of the Theory of Relativity. Part 1. Tensor Calculus.* This book, 167 pages long, was published in 1924 in Leningrad by Academia Publishers, one of the largest and most authoritative publishing houses at that time. According to V. A. Fock (and some other contemporaries), the major role in writing the book was played by Friedmann, who, as a mathematician, was naturally to do the bulk of the work in the first (and planned second) "mathematical" parts. This book is an outstanding introduction to the mathematical apparatus of the theory of relativity in terms of depth, detail and simplicity. It is just a little outdated in mathematical notation (this was a cause of concern for the authors – the apparatus was developing rapidly, and while the book was being printed, articles came out with a better system of notation), which may present some difficulty for present-day readers.

One can only speculate what this book would have been like, had it been completed, with all its five parts, by the authors. Fortunately, Friedmann's other enterprise, his book *The World as Space and Time*, was fully accomplished.

The book and its period

This small book (only 131 pages) was originally written as an article for the philosophical journal *Mysl* (Thought) and came out as a separate edition in Petrograd in 1923 in the series *Contemporary Culture* put out by

Academia Publishers. The printrun of 3,000 copies was normal for the time, and it was "printed in the Military Printing Shop of the Head-quarters of the RKKA [Workers' and Peasants' Red Army] (10 Uritsky Square)."

Today this book can only be found in the largest libraries, and has long become a bibliographic rarity. More accessible to the modern reader is the second edition, published 42 years later, in Moscow, by Nauka Publishers as part of *The Popular Science Series of the USSR Academy of Sciences*. Its printrun this time was 45,000 copies. In this edition, the publisher's preface says quite correctly that "A. A. Friedmann's book has not lost its value. It will be of interest and use to all students of modern physics and its history." But the reader can be misled by the following words: "This is the first popular presentation of the theory of relativity in the Russian language." In actual fact, there had been quite a few popular science books on relativity (both special and general) before Friedmann's book.

Title page of *The World as Space and Time* by A. Friedmann.

This is, incidentally, mentioned by the author himself in his introduction to the book.

The third edition of the book came out a year later, in 1966, as part of the already-mentioned collection: A. A. Friedmann, *Selected Works*. Friedmann's book is not remarkable because it contains a popular account. On the contrary, from the very beginning the author denies its belonging to the "popular" genre: "The presentation of this article," he writes in the very first pages, "by no means claims to be popular and requires, which is absolutely necessary for the essence of the matter, a good knowledge of at least some elements of higher mathematics." Friedmann contrasts his paper with "various attempts to popularize the relativity principle, which it is impossible to popularize; this popularization is achieved either at the cost of totally obscuring the ideas underlying the relativity principle or, what is perhaps even worse, at the cost of perverting its ideas." He adamantly opposes the vulgarization of science, the superficial infatuation with relativity: "Owing to the 'fashion' for the relativity principle and the large number of popular books and lectures dealing with it, distorted ideas have spread not only among the public, but also in specialist quarters, about the finiteness, closedness, curvature and other properties of our space, which are supposedly established by the relativity principle."

When Friedmann says it is "impossible to popularize," he is warning the reader against the temptation to learn about the object of "fashion" without effort. He seemed to be irritated by the fact that for "the public" it was, in his opinion, only a new fashionable topic, a kind of "fruit of education."[1] This is what the book *Fundamentals of Relativity Theory*, which was written at the same time, says:

No physical theory has been, to the same degree as the general relativity principle, the object, on the one hand of admiration and worship, and, on the other hand of indignation and persecution, not only on the part of physicists, astronomers and mathematicians equipped to judge it, but also on the part of people lacking knowledge in these sciences . . . Relativity theory is described not only in a huge amount of scientific and philosophical literature, and also popular scientific literature, but even in newspaper and magazine articles, quite often of low quality, which has created for it a purely superficial success – a kind of fashion, as a result of which the relativity principle is shown in the cinema, and almost anyone believes he has the right to judge some of the most complex and difficult issues in physics.

---

[1] As it was satirically described by Leo Tolstoy in his drama with the same title many years before.

And yet, this public interest in the theory of relativity, in a country which had just gone through gigantic upheavals, which was only beginning to restore life after devastation and hunger, is an amazing phenomenon which cannot be attributed solely to the fads of "society." Light-minded fashion apart, there seemed to be an emerging and growing attention given by the broad mass of readers and audiences to the "eternal" questions of the structure of the world, and the place of mankind within it. It was not without reason that public lectures on relativity theory attracted such large audiences, and the books and articles about it, despite the printruns, quite large for the time, were quickly sold out. Friedmann's works about the great number of popular books and lectures, although said in a definitely critical context, are at the same time remarkable evidence from the most authoritative contemporary about the attitudes and interests of those days.

"Impossible to popularize"

The library of the USSR Academy of Sciences in Leningrad, one of the finest collections of scientific literature, does not have very many popular science books and pamphlets on general relativity published before 1923. There are only eight such books in Russian there. In fact, many more existed – we are aware of at least twenty-odd titles. About half of them are translations of foreign, mainly German, books into Russian. First of all comes the world's most famous book in physics – Einstein's book *Relativity: the Special and the General Theory. A Popular Exposition*, published at that time in Russian in several successive editions. The Petrograd editions of 1921 and 1922 were translations of the tenth German edition made by Sergei I. Vavilov, the future President of the USSR Academy of Sciences. In 1924 Felix Auerbach's *Space and Time* came out, also in Vavilov's translation. The book, in Vavilov's words, has "the usual virtues of this author's works: it is easy to understand and simple and absorbing in presentation." No less successful was E. Freundlich's *Fundamentals of Einstein's Gravitation Theory*, with a favorable foreword by Einstein. The translation was edited by Frederiks. Mention should also be made of I. Lehmann's *Relativity Theory* and E. Cassirer's *Einstein's Relativity Theory*, both published in Russian translation in 1922. According to the authors, the manuscripts of both these books had been read and approved by Einstein. Original books, as opposed to translations, comprised mostly slim pamphlets. Such, for instance, was Professor S. Lifshitz's book *A. Einstein's Relativity Principle* (Moscow,

1922) which, as its cover says, was a lecture read "at a joint meeting of all students' scientific groups at Moscow University."

It is interesting to note that there were about two dozen books on relativity theory in V. I. Lenin's library in the Kremlin. Among them were the above-mentioned book by Einstein in a German edition of 1921 and in two Russian editions of 1922 and 1923, M. Born's popular science book *Relativity Theory* in a Russian edition of 1922, two books by C. Nordmann about Einstein in Russian translations of 1922 and 1923, books by O. D. Khvolson, A. V. Vasiliev, E. Cassirer, N. A. Morozov, A. Mashkovsky, V. V. Arnold, H. Schmidt, H. Bergson, Ye. S. London, and two collections of articles on modern physics and its philosophical questions.

In the catalogue of Lenin's library, published in 1961, item No. 4598 on page 405 is A. A. Friedmann's book *The World as Space and Time*.

The 1916 German edition of Einstein's book was also in Friedmann's personal library, which is now in the Main Geophysical Observatory. It has mostly scientific editions – books on relativity theory by M. Laue, W. Pauli, and Ya. I. Frenkel. Friedmann's library has a total of about 900 books in mathematics, mechanics, physics, meteorology, sea navigation, aviation, etc.

It is hard to say what Friedmann thought about each of these books; but he did not think highly of popular science literature as a whole, as is clear from his above-quoted words. His own book was not that of a popularizer, even a highly competent, skilled and experienced one. The book was the product of a creative effort by one of the founders of modern physics. It is on a par with Einstein's book; it continues the latter, but also sometimes argues with it. The argument with Einstein was started, though, not by this book, but by Friedmann's first cosmological study published the year before. In the book one hears only a few echoes of this great dispute. Yet, one is clearly aware of a fresh perspective on the subject, and a new and thoroughly thought-out way of presentation.

Friedmann chooses a non-traditional approach which takes the reader right to the essence of the matter. This concerns first of all the attitude toward mathematics. According to Friedmann, the mathematical apparatus of general relativity is formidable, and this would present an insurmountable obstacle to lay readers, if they tried to read the scientific literature without preparation. The author helps the reader to overcome the mathematical barrier. What is remarkable is that he is leading the reader toward the goal not by avoiding his apparatus (as was done by Einstein and others after him), but with the help of mathematics. Friedmann selects and explains thoroughly the key concepts of geometry, which is the mathematical tool of general relativity. And then it turns out that mathematics is not an obstacle, but the direct way toward the goal, if only one takes the trouble to understand the formulae.

The new approach was not restricted to mathematics. It is well known that, historically, general relativity appeared after the special theory was developed in 1905. Both in popular and in "serious" books its presentation always followed that of special relativity. This was done by Einstein and others after him. Friedmann moves away from the established tradition, abandoning the historical method of presentation and choosing the logical approach. He immediately introduces the reader to the world of general relativity: "Logically, the special relativity principle, without much damage to the essence, can be left almost wholly aside. This is exactly what I am going to do." The special theory is a particular case of the general theory, being valid provided that the gravitational fields are weak. Logically (and ideologically) the general theory is not only broader, but also, so to speak, more immediate: it bases itself on the simplest initial facts and ideas.

These two features, one mathematical, and the other logical, or to be more exact, physical, place Friedmann's book somewhat outside the traditional genres: it does not belong to popular science literature, which, in Friedmann's words, "explains nothing," but it is not a "scientific" book either, which would require "a formidable mathematical apparatus for its correct understanding." He would like to make his book generally comprehensible, although he preferred not to call it popular, so as not to create a too easy attitude in the reader: "so as to avoid a wrong conception of the present article in the reader's mind . . ." Making this warning, Friedmann hopes that the reader will take a serious attitude toward the subject. And for the interested reader he solved brilliantly the "unsolvable" task of popularization. His book is a not too easy, but deeply engrossing read. "The grandiose and bold range of thought characterizing the general concepts and ideas of the relativity principle" is conveyed in the book with great inspiration, in a convincing, clear and simple way.

Friedmann's work has fully retained its purpose and significance down to our own times. To today's reader it gives, among other things, the opportunity to feel the book's subject and theme as it was experienced by the author, one of the most outstanding Russian scientists, a classic of world science, a founder of the modern science of the Universe.

We whole-heartedly recommend this book to the reader.[2] Here we shall

2 Unfortuntely it is not available in English.

only give some of its passages, arguments, explanations and examples. It would hardly be reasonable or possible to give a short summary of the book from beginning to end. Friedmann's style is rigorous, logical and concise. There is virtually nothing secondary, and the reasoning is constructed as a chain in which every link is necessary, so that it cannot be abridged without breaking the connections. It is a highly condensed introduction to the complex and deep science of the world, space, and time.

Space

The first chapter of the book is devoted entirely to the theory of space's properties. It is in fact called "Space." What is meant is space in a very broad meaning of the word. Later, the geometrical theory will be applied to the world understood as a four-dimensional unity of space and time. The geometry necessary for general relativity is not the "conventional" Euclidean geometry which we are all familiar with, but a broader mathematical theory which incorporates, of course, the "usual" geometry, but only as a simplest particular case.

Friedmann is a mathematician, but he applies his science to the real world and therefore acts as a physicist, as a natural scientist. "We accept, as something ready-made, the interpretation of geometry which physicists are used to," he writes, explaining: "Thus, for instance, a very small material body represents (approximately) a point in a geometrical space; a light ray represents a straight line, etc." In the very first pages of the book he speaks about the task of studying experimentally the properties of physical space. But to make this possible one should develop a conceptual basis for such studies, create a system of notions in terms of which one can give an exact formulation of the problem, analyze it theoretically and draw up a program of necessary experiments and observations. This is what mathematics is needed for. What is needed is the kind of geometry which generalizes the "conventional" Euclidean geometry and which, fortunately, mathematicians (Lobachevsky, Gauss, Bolyai, Riemann) had already created by the end of the 19th century independently of physicists and their interests. The accepted name for this geometry is Riemannian geometry; it deals with a space whose properties, in the most general case, are different at different points.

Following David Hilbert,[3] his authority in the field of mathematics and relativity theory, Friedmann gives this definition of space:

[3] The "godfather" of Friedmann's first scientific article.

Geometrical space, or simply space, is a set of entities called points, lines, surfaces, angles, distances, etc. which are in certain relations to each other, established by a system of axioms and theorems following from these axioms.

A point is the simplest of the above-mentioned "entities." If a space has three dimensions (as is the case with the "conventional" Euclidean space, e.g. a room or a house), each point of such a space can be assigned three numbers. And this can be done in such a way that one and only one point of the space will correspond to each triplet of numbers. This procedure "will be called, for short, the arithmetization of space."

"Arithmetization" is a term which is no longer frequently used. Friedmann uses it quite often in both of his books on relativity. It appears that he liked a high degree of generality in the notion of arithmetization: it is a way of establishing rules for the numerical description of any properties of any entities. Here is the simplest case of arithmetization given by Friedmann. "Humans have the property of gender; let us assign the number 0 to the property of being female, and the number 1 to the property of being male; the establishment of this rule is the arithmetization of the given class of properties."

Pelageya Yakovlevna Polubarinova-Kochina, Friedmann's pupil, and now a Full Member of the Academy of Sciences, recalls that Friedmann's women colleagues didn't like this example very much, and "women workers of the department expressed their indignation to Alexander Alexandrovich . . . He promised to make it the other way round in the second edition."

Upon arithmetization a label is put against each "point" of physical space with three numbers corresponding to this point written on the label. We are so much used to this arithmetization of physical space that we do not give it any thought at all and do not see either its essence or its arbitrary character. Meanwhile, the names or the numbers of streets, houses, floors, apartments is a case of arithmetization of physical space which is astonishingly arbitrary (particularly in our provincial centers and in Moscow).

The author stresses again and again: "The process of arithmetization of space is absolutely arbitrary; the choice of the above-mentioned triplet of numbers is not predetermined by anything . . . Triplets of numbers by means of which we arithmetize space are called coordinates." Since the choice of coordinates is "absolutely arbitrary," different ways of assigning various triplets of numbers to points are possible. "Switching from one method of arithmetization of space to another is called a transformation of coordinates."

The introduction of coordinates of points as such is well known, for example, in planimetry – Euclidean geometry in the plane. In this case we have a two-dimensional space, and each point is assigned not three, but a pair of numbers. In the plane one can introduce rectangular (Cartesian) coordinates, or polar coordinates, etc. None of these coordinate systems is in any way better or worse than another. But what is important is that in the case of an Euclidean plane the properties of the space itself are known to us in advance. And we construct our system of coordinates taking these properties into account.

Yet, in the most general, non-Euclidean space (no matter whether two-, three- or four-dimensional) the introduction of coordinates does not require or presuppose knowing the properties of space in advance. Rather, it is to make the description of the properties of space possible that coordinates are introduced; with the help of coordinates these properties can be precisely formulated and then studied.

Properties of space can be divided into two types. Properties of one type are wholly dependent on the chosen (always arbitrary) system of coordinates, whereas others will remain unchanged, no matter what coordinates we use. These latter are evidently the most important, because they characterize space as such, irrespective of the coordinates chosen by one means or another. Properties of the second type can be expressed through propositions whose form does not change with transition from one system of coordinates to another. One calls such properties invariant relative to a transformation of coordinates.

Here is a simple example of invariant properties of space. In a space, one can consider points, lines and surfaces. If we say that a line contains a given point of space (i.e. passes through it), this proposition is true under any system of coordinates. Or, if we say that a surface contains a certain line (i.e. the line is "drawn" across the surface), this proposition does not depend on the coordinates either. The property that a line contains a given point and the property that a surface contains a given line are invariant properties.

Thus, Friedmann begins to acquaint his reader with the basic notions of geometry. To assimilate them is "absolutely necessary for the essence of the matter."

The metric

In a Euclidean plane one can take two points, connect them by a straight line, and then the segment of the line between these two points will be the distance between them. One can, of course, connect points not only by

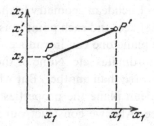

Fig. 1   The distance between points in a plane.

straight lines, but by various curved lines as well, and measure distances between points along arcs of these curved lines. Yet, if one uses a straight line, then the distance between points is known to be the shortest. Everything is simple here: the shortest way is the straight way. It can be agreed to consider the shortest distance along a straight line as the distance between points.

If so, one can easily write several simple formulae which will give a mathematical expression for the distance between points in a plane. For this, one should first choose a coordinate system. Let us choose, for example, the rectangular coordinate system and designate one coordinate as x_1 and the other as x_2. Then, as is seen from Fig. 1, the square of the distance between the points P and P' will be expressed in terms of the given number pairs (x_1, x_2) and (x_1', x_2') as:

$$(P, P')^2 = (x_1' - x_1)^2 + (x_2' - x_2)^2.$$

This formula is Pythagoras' theorem: the square on the hypotenuse (P, P') equals the sum of the squares on the other two sides $(x_1' - x_1)$ and $(x_2' - x_2)$. To obtain the distance, it remains only to take the square root.

The notion of a distance can be introduced not only in a plane, but also in a space, about whose properties nothing is known in advance. One should only keep in mind that among all thinkable (and studied in mathematics) spaces there are some for which the notion of distance does not even exist. But geometries so "poor" in properties are not used in general relativity. The latter uses a geometry of spaces in which the distance between two points always exists. It is said in such cases that the space has a metric.

Distance has two fairly obvious properties:

(1) the distance does not depend upon the order of the points:

$$(P, P') = (P', P);$$

(2) the distance between two coinciding points (points with the same coordinates) equals zero:

$$(P, P) = 0.$$

It turns out that it is most convenient to use not the distance (P, P'), but the square of this quantity $(P, P')^2$, as in the formula we have written above for a plane. For the square of the distance, properties (1) and (2) evidently hold.

These properties are, as we say, obvious. But they are obvious for the simple reason that such properties are well known to us from the "conventional" Euclidean geometry. In actual fact, these are certain constraints imposed on the notion of distance which are introduced without any reference to Euclidean geometry. As Friedmann says, "these constraints will, of course, play the role of axioms in the system of geometry which we are developing."

To these two properties of space is added a third, which again holds in Euclidean geometry; this property concerns the distance between two points which are very close to each other. In this case the differences of coordinates of the points $\Delta x_1 = x'_1 - x_1$, $\Delta x_2 = x'_2 - x_2$ are very small, or infinitesimal, to use the language of differential calculus. So it is stated that the square of the distance between sufficiently close points can be represented as an infinite sum of various powers of these differences times suitable coefficients.

It is easy to see that the most significant terms, i.e. the largest ones in this sum, will be those which contain the differences of coordinates to the lowest powers. How many such significant terms will the expression for the square of distance have in a two-dimensional space? Friedmann gives a simple and clear (although necessarily not very short) proof that there will be three such terms and they will be expressed in the following way:

$$(P, P')^2 = g_{11} \Delta x_1^2 + 2g_{12} \Delta x_1 \Delta x_2 + g_{22} \Delta x_2^2.$$

These are quadratic combinations of the differences of coordinates – their squares and their product. Each of the three combinations has its coefficient g_{11}, g_{12} and g_{22}. The three latter quantities are not, generally speaking, constant, but depend on the location of the two points in question within the two-dimensional space. They are functions of the coordinates of one such point, e.g. P.

The expression for the square of the distance between two infinitely close points is of primary importance in Riemannian geometry. It turns out to contain a description of all the geometric properties of any infinitesimal region in the Riemannian space. These properties are expressed by the functions g_{11}, g_{12} and g_{22}.

To understand the meaning and role of these functions, let us conduct a thought experiment for the purpose of finding the values of these functions at a certain point of some non-Euclidean two-dimensional space. Here is the experiment suggested by Friedmann.

Let us assume that a number of curves are drawn through a given physical point P, and a point P' is taken very close to P on each of the curves; measuring the physical length of the arcs along each of the curves between P and P', we shall obtain a series of numbers (P, P'). In addition, we shall need to know the differences between the coordinates, Δx_1 and Δx_2. This will enable us to obtain a set of equations similar in form to the last of the above expressions for $(P, P')^2$. In such equations the lengths of the arcs (P, P') and the differences of coordinates will now be known quantities, while the quantities g_{11}, g_{12} and g_{22} will remain unknown; "by making a sufficient number of measurements we will be able to obtain a sufficient number of equations from which to determine" the sought-for quantities g_{11}, g_{12}, and g_{22}.

In a two-dimensional space this will require at least three measurements. Having made three measurements, we shall obtain three simultaneous equations which can always be solved (provided only the points do not coincide with each other). If we made such measurements in an Euclidean plane and used rectangular coordinates we would obtain the following results:

$$g_{11} = 1, \qquad g_{12} = 0, \qquad g_{22} = 1.$$

As Friedmann observes here, "what has just been described with sufficient pedantry took place in Egypt where practical life and religious needs established the foundation for our Euclidean metric."

While working with our geometrical formulae, one should make one more transformation and change of notation. Instead of the differences Δx_1 and Δx_2 Friedmann introduces infinitesimal increments of coordinates, differentials dx_1 and dx_2. The distance (P, P') will be further denoted by ds. Then, limiting ourselves, as above, to the most significant terms in the sum expression ds^2 in terms of dx_1 and dx_2 we can write

$$ds^2 = g_{11}dx_1^2 + 2g_{12}dx_1dx_2 + g_{22}dx_2^2.$$

The set of quantities g_{11}, g_{12} and g_{22} is called a metric tensor. These quantities can be all designated as g_{ik}, if we remember that i and k take the values 1 and 2, and $g_{12} = g_{21}$.

In the particular case of a Euclidean plane when the metric tensor takes the values $g_{11} = 1$, $g_{12} = g_{21} = 0$, $g_{22} = 1$, the square of the distance between infinitely close points has the form

$$ds^2 = dx_1^2 + dx_2^2,$$

i.e. we again come back to Pythagoras's theorem, but only for a triangle with infinitesimal sides.

If one makes measurements on the Earth as described above, and with an accuracy much higher than in ancient Egypt, the result, as we understand it, will reveal that real geometry does not conform to Euclidean geometry. We live not on a plane, but on a curved surface which is fairly close to a spherical surface. A sphere is an example of a non-Euclidean two-dimensional space.

Small segments of a sphere differ little, of course, from segments of a plane, and the less the width of such segments compared with the radius of the sphere, the less is the difference. Plots of land which ancient land-surveyors (geometers) had to deal with were small, and therefore, in keeping with their view of the geometry of the Earth's surface, the most reasonable and quite sufficient approximation was the Euclidean plane approximation.

If spherical surface segments are not too small, deviations from the plane become conspicuous. We shall now see how the non-Euclidean geometry of a sphere is described.

It is convenient to study the metric of a sphere using not rectangular, but angular coordinates of latitude and longitude, known from geography. To be more precise, let us introduce the angle ϑ as x_1, which is 90° minus the latitude, and the longitude φ as x_2. Then measuring, with appropriate accuracy, the length of the arcs going through the given point on the surface of the sphere, we shall obtain:

$$g_{11} = R^2, \qquad g_{12} = 0, \qquad g_{22} = R^2 \sin^2 \vartheta.$$

The radius R of the sphere appears in these formulae. Both non-zero components of the metric tensor are proportional to the squared radius. One of them, g_{11}, simply equals the squared radius, while the other one, g_{22}, contains also the squared sine of the angle ϑ. This latter quantity is different at different points of the sphere's surface, so that we have here a dependence of the metric tensor defined at the given point, on the coordinates of this point, which is provided for by the general theory. In order to be able to guess that there should be the squared sine here, one had, of course, to make a series of additional measurements at different points of the sphere with different latitudes.

The squared distance between infinitely close points on a sphere is represented in terms of the coordinates ϑ and φ in this form:

$$ds^2 = R^2 d\vartheta^2 + R^2 \sin^2\vartheta \; d\varphi^2.$$

This formula is also, as a matter of fact, Pythagoras' theorem: if one draws on a sphere a rectangular triangle with the small sides $Rd\vartheta$ and $R\sin\vartheta \; d\varphi$, the distance ds will be the hypotenuse of such a triangle.

We have dealt so far with two-dimensional spaces. When one goes over to three-dimensional spaces, the number of quantities making up the metric tensor increases to six. These will be quantities g_{ik} depending on three variables x_1, x_2, x_3, and here i and k vary from 1 to 2 to 3. For the squared distance between infinitely close points one will, by analogy, obtain:

$$ds^2 = g_{11} dx_1^2 + g_{22} dx_2^2 + g_{33} dx_3^2 \\ + 2g_{12} dx_1 dx_2 + 2g_{13} dx_1 dx_3 + 2g_{23} dx_2 dx_3.$$

In the ordinary three-dimensional Euclidean space

$$g_{11} = g_{22} = g_{33} = 1, \qquad g_{12} = g_{13} = g_{23} = 0.$$

And then the squared distance is expressed through the "three-dimensional" Pythagoras theorem:

$$ds^2 = dx_1^2 + dx_2^2 + dx_3^2.$$

Finally, the last thing that should be said about the space metric: the squared distance between two infinitely close points is an invariant quantity: the value of this distance is the same (in the same units) for any, however different, systems of coordinates. The squared distance expresses the specific geometric properties of space.

Curvature

Starting the presentation of one of the most difficult geometrical notions, namely the notion of space curvature, I believe it necessary to warn that this paragraph will outline only the most elementary bases of the theory of curvature, since a deeper acquaintance with the theory of space curvature requires a much larger mathematical apparatus than the one which we are using in the present article.

Friedmann introduces the notion of curvature through the procedure of a parallel vector transfer. In this he follows the Italian mathematician Tullio Levi-Civita (1873–1941) who was a Foreign Member of the St. Petersburg Academy of Sciences, and later of the USSR Academy of Sciences. Friedmann corresponded with Levi-Civita in the 1920s; in 1923, they met in Delft at the Mechanics Congress, and in one of his letters written from abroad to Steklov Friedmann mentions Levi-Civita's interest in his work and that of his associates (see Chapter 10).

Friedmann explains in detail the procedure of a parallel transfer. He begins by explaining a number of necessary concepts: a straight line, an angle, a vector. We shall not reproduce now the exact definitions and the whole process of reasoning. We shall only stress, together with Friedmann, that "the notion of parallelism plays an outstanding role in spaces differing in their metric properties from Euclidean space." For plane geometry the propositions "two straight lines are parallel" and "two straight lines do not intersect" have the same meaning. But this is not the case, for example, on a sphere. The role of straight lines, in the case of a sphere, is played by the arcs of great circles, since it is along great circles that the distance between points on a sphere is the shortest. But two "straight lines" on a sphere always intersect; therefore, in this case the above-given propositions on parallelism and non-intersection do not mean the same thing. Thus, the curvature of the sphere's surface, its non-Euclidean character, is manifested.

In the case of a vector parallel transfer along a closed contour in a plane, the vector returns to the initial point retaining its original direction. But the property of parallelism for a sphere is such that, having performed a parallel transfer around a closed contour, the vector returns to the initial point having a different direction. A non-zero angle between the initial direction and the new direction is evidence of curvature.

Another evidence of curvature is the property of the angles of a spherical triangle. If one constructs a triangle on a sphere from arcs of great circles, the sum of its angles will exceed two right angles. The difference between the sum of the angles and two right angles is called the spherical excess, and it equals, as is known from spherical geometry, the ratio of the area of the spherical triangle to the square of the sphere's radius.

Friedmann uses this geometrical fact to explain what the curvature of a sphere is. If a parallel vector transfer is carried out on a sphere along the sides of a spherical triangle, the angle between the vector's initial and final (after its return) directions turns out to be equal to the spherical excess for this triangle. The measure for the sphere's curvature is the ratio of the spherical excess to the area of the triangle, which, in view of what has been said, equals the reciprocal square of the sphere's radius. It is this quantity which is called the sphere's curvature.

Curvature has the dimension of reciprocal length squared, so its numerical value is expressed, for instance, in cm^{-2} or m^{-2}. There is no arbitrary choice here, except the choice of a unit of measurement. Curvature is a proper, invariant property of space; therefore its quantitative measure does not depend upon the system of coordinates chosen. If, upon measurements and calculations within one system of coordinates, the curvature turns out to be equal to zero, it will be equal to zero in any other system of coordinates. In the same way, non-zero curvature remains non-zero under any transformation of coordinates.

Curvature can be measured if one knows the space metric tensor. Although the type of a metric tensor depends very strongly on what system of coordinates is used, the curvature calculated from it is free from such dependence.

Apart from curvature, geometry also uses the radius of curvature of space at any given point. For a sphere, the radius of curvature is the same for all points and equals simply its radius.

It remains only to add that "throughout our discussion of space, we referred to a two-dimensional space merely for the sake of illustration. It is, however, not difficult to extend the definitions of our concepts and their properties to a space of any number of dimensions; formally, we shall not meet with any difficulties, except an increase in the number of g_{ik} for the fundamental metric tensor." There were three such quantities in a two-dimensional space and six in a three-dimensional one; "finally, in a four-dimensional space the number of quantities required to define the fundamental metric tensor will increase to ten (the number of different pair combinations of the symbols 1, 2, 3 and 4)." Speaking about a four-dimensional space, Friedmann means by the fourth dimension "time, without which there is no space, and which determines not physical three-dimensional space, but physical four-dimensional space – the world."

Thus ends the chapter on geometry in the book *The World as Space and Time*. Friedmann uses as the epigraph to the chapter a quotation from the apocryphal Wisdom of Solomon: "Thou hast arranged all things by measure and number" And the following Chapter 2, "Time and the World," opens with an epigraph from Aristotle: "Time is a measure of motion."

Time

"Our conceptions of space are much clearer than our conceptions of time," says Friedmann. And this does not apply only to the conception which existed in "old" classical physics. Both the special and even the general theory of relativity, having explained much about the nature of time (particularly concerning its mutual relation with space), still failed to give a full and comprehensive answer to the question that had long preoccupied people: what is time?

One could, as Friedmann writes, try and approach the problem of time in the same way as the problem of space: "We began the chapter on space with a definition of abstract space; we could do the same with time, too, saying that time is a set of entities called instants which are in certain relations between each other and space. I would rather follow a different course and begin by considering physical time."

Having chosen a physical approach, Friedmann follows Aristotle's tradition and connects time with motion, with the movement of physical bodies in space. But at the same time from the very start he wants to break one centuries-old habit. He explains to the reader that the time reckoning which we are used to and have long considered the only one possible, universal and truthful, is in actual fact an arbitrarily chosen measure of duration which got established for practical and psychological reasons. This is the tradition of time reckoning using our planet Earth as the clock. But this is by no means the only way to reckon time.

Friedmann invites the reader to conduct several thought experiments. The first consists in observing the motion of the end of a pointer of a fixed length, which is directed from the Earth's center toward some star. The time shown by such a pointer – and it rotates, of course, together with the Earth's daily rotation around its axis – can, it is suggested, be called star time for short (Friedmann warns that astronomers use the term "sidereal time," i.e. star time, in a different sense). This is "a very convenient time, since a great many motions will take place in such a way that the length of the arc traveled between two points on the trajectory will be proportional to the difference of star times:" $s = vt$. Here s is the length of the arc, i.e. the path traveled by a body along its trajectory during a time interval t; and v is a constant which, as is well known, has the meaning of velocity. "We shall call such motions uniform with respect to star time."

And again: "I repeat that a very great number of motions will be either uniform or virtually uniform with respect to star time. Therefore, the investigation of motion is considerably simplified when it is star time that is chosen to represent time."

Here is what Friedmann says about the origin and meaning of such time reckoning:

It is exactly such time that was first chosen by mankind, and the choice of this very time was historically, of course, quite inevitable and natural, if only because of the immense religious and mystical impression which star time and its stately motion produced and even now produces on the human soul. The choice of star time as universal time has had a tremendous impact on the whole history of culture and, of course, on the formulation of the principal laws of mechanics . . . This course, which from our point of view is absolutely arbitrary, seemed the only true and sacred one, and star time, ascribed a number of mystical properties, became time as such, enigmatic and difficult to comprehend.

What other time can one conceive of, besides star time? Friedmann says that one should not choose one and only one kind of motion as the time capable of moving a clock hand, not even the Earth's daily rotation, which determines the natural rhythm of our earthly affairs. Absolutely any movement can be taken as such a motion. There only needs to be a body moving along its trajectory, and then one can measure time segments along the arc of this trajectory. Such motion can be either uniform or non-uniform with respect to star time. Here is a specific example suggested by Friedmann – one more thought experiment:

Of course, not every motion is uniform with respect to star time; the free fall of bodies near the Earth's surface, or rather in a constant gravitational field, is an example of non-uniform (uniformly accelerated) motion with respect to star time. The lengths of an arc traveled by a point in this motion is expressed in terms of star time by the following formula:

$$s = at^2 + b,$$

where a and b are constants, with a depending upon the free fall acceleration: $a = g/2$, where g is the free fall acceleration in our constant gravitational field.

This uniformly accelerated motion can also be used to reckon time. Time measured by this motion "we shall call gravitational, for short."

Let us assume we have a clock with a hand whose end describes an arc of length $s = at^2 + b$. Then t is star time, and s is gravitational time. It is not very important, of course, that the quantity s represents length, not time; as a matter of fact, units of time measurement are arbitrary, too: they have been agreed by tradition and for convenience' sake. Time can quite well be measured not only in seconds, hours or years, but also in centimeters, parsecs, grams, electron volts, etc. One need only know the way units of one type are converted into another, and then there will be no misunderstanding.

If one marks star time with an asterisk, $t = t_*$, and gravitational time with the symbol g, $t_g = s$, the relation between the two times can be expressed by these formulae:

$$t_* = \sqrt{[(t_g - b)/a]}, \qquad t_g = at_*^2 + b.$$

With respect to star time a stone falls non-uniformly, and with respect to gravitational time, stars move non-uniformly. In terms of gravitational time the hands of all our clocks and watches move non-uniformly, repeating the motion of the stars, i.e. the Earth's daily rotation; but a heavy projectile falls uniformly.

Obviously, the author does not suggest that we should change from traditional star time to gravitational time.

It is natural that if we adopted gravitational time as true time we would have to transform radically the whole of mechanics. This change would make mechanics much more complicated, since a number of the simplest and most frequently occurring motions would turn out to be non-uniform motions, following fairly complicated laws. The introduction of gravitational time would, of course, be inexpedient, but, except for the principle of expendiency and convenience, gravitational time would have the same right to serve as time, as star time has.

Friedmann finds it relevant to note that the idea of the arbitrariness of time reckoning was established by St. Augustine – a philosopher and Christian theologian (354–430), who reflected much on the nature of time and its perception by mankind. Friedmann cites the following passage from St. Augustine:

I heard a learned man once deliver it, that the motions of the Sun, Moon and stars, were the very true times, and I did not agree. For why should not the motions of all bodies in general rather be times? Or if the lights of heaven should cease, and the potter's wheel run round; should there be no time by which we might measure those whirlings about . . .?

This thought, expressed, as Friedmann says, in "amazing words" many centuries ago, remains true today.

Thus, time – "a measure of motion" – can be reckoned through any pre-chosen motion. The transition from one way of reckoning time to another, i.e. from one clock to another, is similar to a transformation of coordinates, when we will change from one coordinate system to another. The substitution of one time for another will, of course, be carried out differently at different points of space. As Friedmann notes, this last fact indicates a much greater closeness of time to space than is commonly accepted.

Time, like space, has properties which are not intrinsic, i.e. which depend on the method of time reckoning. Thus the time interval between two events, even if they occur at the same point of space, is not an invariant quantity. This not very obvious property of time has been discovered by general relativity; this theory has deeply and fully reflected the features of time which unite it with space and make it the fourth coordinate of the world.

A four-dimensional world

"Henceforth space by itself, and time by itself, are doomed to fade away into mere shadows, and only a kind of union of the two will preserve an independent reality." Friedmann took as one of his epigraphs these famous words of Hermann Minkowski, a great German mathematician. They were said concerning the world of special relativity. The world of general relativity is a single four-dimensional space–time, the properties of which are, in fact, different at different points and at different moments of time. "Four-dimensional points" in space–time are, to use the modern term, events (Friedmann preferred to call them phenomena), each of which is characterized by four numbers. These numbers are assigned to events in accordance with the particular arithmetization of space–time adopted, i.e. a system of four coordinates – three of space and one of time. As is the case with "conventional" space, space–time has its inherent properties not depending on the system of coordinates chosen. These properties are characterized by four-dimensional invariants. The most important of them is the interval, i.e. the four-dimensional span between two point-events infinitesimally close to each other.

The mathematical expression for an interval is derived similarly to the expression for the distance between two infinitesimally close points in a three-dimensional space. Simply, the three space coordinates x_1, x_2, x_3 are complemented by a fourth, the time coordinate x_4, which has an equal standing with the first three:

$$ds^2 = g_{11}dx_1^2 + g_{22}dx_2^2 + g_{33}dx_3^2 + g_{44}dx_4^2$$
$$+ 2g_{12}dx_1 dx_2 + 2g_{13}dx_1 dx_3 + 2g_{14}dx_1 dx_4$$
$$+ 2g_{23}dx_2 dx_3 + 2g_{24}dx_2 dx_4 + 2g_{34}dx_3 dx_4.$$

Apart from the order of the terms (which is, of course, unimportant), time enters this expression in the same way as any of the three space coordinates. It is regarded here as having the same nature as space. Friedmann writes:

> Time does not differ in any way from the other coordinates.
> Is this last conclusion correct? A more careful consideration of the question will show that the last conclusion cannot be recognized as completely correct . . . I cannot dwell on this detailed question which is relatively little-developed in the Theory of Relativity. I will only note that the reason for restoring to time its outstanding significance is the causality principle, one of whose requirements is that one must'not change the arithmetization of the physical world in such a way that cause and effect change places.

Indeed, time has something which can in no way be reduced to space correspondences and has no analogies in geometry. This is first of all its irrepressible flow. Time flows, and always in one direction – from past to future. Its flow is such that cause always precedes effect, and therefore, moments of time in the history of each physical body cannot be transposed: the order of their sequence is unchangeable. The flow of time is a special property of the physical world, obvious to everybody, but even now remaining insufficiently understood and studied. (For various present-day attempts to solve the enigma of time see A. D. Chernin's book *Fizika Vremeni* (The Physics of Time), Nauka Publishers, Moscow, 1987).

Let's go back to our expression for an interval. It contains ten independent quantities g_{ik}, comprising the metric tensor of a four-dimensional space–time. As in the case of space proper, a question naturally arises on the experimental definition of the metric. "Measuring intervals in the physical world will make it possible for us to study experimentally the metric of the world," Friedmann writes. But, when measuring the interval,

> we immediately come up against the fact that to dispose ourselves arbitrarily in time is beyond our power; thus we can determine the metric of the world only over a very limited time.
> Further, the possibility, in principle, of studying the geometric world is hampered by an essential defect of our experimental equipment, namely in difficulties in measuring the interval. We need to bypass in some way or another both above-mentioned difficulties and find quantities which could be measured by our technical means throughout the world.

It is a task at which physicists and astronomers are still working. Yet, so far there has been little progress in this direction. Far more successful have been observational studies, not of the geometry of the world, but of its dynamics. It is on this path that Friedmann's remarkable theoretical discoveries in cosmology found support. We shall deal with this in more detail later in this book.

It has been said above that the interval between two events is an invariant quantity. This is a quite natural property of an interval, which is just the distance between two points in a four-dimensional world. It is well known that in conventional Euclidean space the distance between two points does not depend on the system of coordinates in which it is calculated. For instance, in a plane, a segment of a straight line connecting two points has a definite length, no matter in which coordinates – rectangular, oblique, polar or any others – we measure it. This property of distance remains valid in the four-dimensional world too.

One property of time which we mentioned a little earlier (in the preceding section) also derives from the invariance of an interval. If one considers the interval between two events taking place at a single point of space, the differentials of the space coordinates in the mathematical expression for this interval will be zero: since both events took place at the same point the space coordinates did not change. Consequently, the distance between two events is determined in this case only by the distance in time. But in a different reference system the time interval between the same events will be different.

If there is a physical body and we impose on it a system of four coordinates, i.e. place on this body a clock and the origin of space coordinates, the clock will show the so-called proper time of this body. Intervals of proper time between different events in the history of the given body are expressed, according to the above, in invariant quantities.

World equations

Riemann used to say that the geometry of the real world appears to be determined by physical phenomena taking place in time and space. An epigraph taken from Riemann opens the third and last chapter of Friedmann's book, which is entitled "Gravity and matter." The establishment of a link between geometry and physics is the main accomplishment of Einstein's general theory of relativity.

The metric tensor at different points of space and its changes in time are determined, according to Einstein, by the location and motion of physical bodies – of all the matter in the Universe. The metric of the Universe thus turns out to be linked with physical processes taking place in the world. Space and time have thus completely lost the absolute nature ascribed to them by classical physics. Now, according to the new understanding, they constitute a single four-dimensional space–time, whose geometrical properties can differ, to any extent, from Euclidean ones. Deviations from Euclidean properties are the greater, the higher is the density of the matter, the more its mass and energy are concentrated in the given region of the world. The link between matter and geometry in the world is forged by gravitation, which, since Newton, has been called universal. Gravitation differs from other fundamental physical interactions in that all bodies in Nature are subject to it. In the 1920s the only known interaction other than gravitation was electromagnetism; later they were supplemented by the discovery, in the physics of atomic nuclei and elementary particles, of two further interactions: the strong and the weak. These manifest themselves only at very small distances, comparable to the sizes of elementary particles (for instance, the proton), and in no way can be considered universal: on the scale of bodies of "ordinary" size and even more so on the scale of the Universe their action is negligible. Electromagnetism cannot be considered universal either. Although it has no direct limitations as to spatial scale, electromagnetic forces operate only when there are electric charges or currents.

The universal character of gravitation, to which all natural bodies are subject, determined its key role in the general theory of relativity. Establishing the relations between geometry and physics, this theory has at the same time revealed the nature of gravitation itself, which had always been a mystery. Newton discovered universal gravitation, and gave it a mathematical formulation: two bodies are attracted to each other with a force proportional to the product of their masses and inversely proportional to the square of the distance between them. But he gave up the attempt to explain the intrinsic nature of gravitation; for this, as he himself wrote in the *Principia* (1687), one should know far more facts than those which were then available, and "I frame no hypothesis."

"In explaining the nature of gravitation," writes Friedmann, "a major role, at least historically, was played by the fact, proved experimentally with great accuracy, that the mass which is part of the formulation of the laws of universal gravitation (gravitational mass), coincides with the mass mentioned in Newton's second law (inertial mass), according to which mass times acceleration equals force . . . If this circumstance (well-known to Galileo and Newton) is not accidental, one naturally wishes to explain gravitational force kinematically and consider it as an imaginary force or (in accordance with d'Alembert's principle) an inertial force." If one accepts this point of view (which Einstein did) it becomes possible to link "the world metric with the phenomenon of gravitation and thus include astronomy and celestial mechanics, with their strikingly precise methods, among the topics concerned with the experimental establishment of the world's geometry." Of decisive importance here was the hypothesis (one could not do without hypotheses after all) that "a material point under the action of gravitating masses moves by inertia. It opened the path which leads to the establishment of the main propositions of Einstein's general principle of relativity."

Thus, gravitating masses distributed and moving in world space constitute the geometry of the world. They may be either individual bodies moving in the void or a continuous medium similar to a gas or fluid. It is through the characteristics of matter – density, pressure, velocities – that the metric of the four-dimensional world is determined. There exist "world equations" that link the density, pressure and velocites of matter with the metric tensor.

Friedmann does not give in his book the explicit form of these equations: "The establishment of world equations and even their simple explanation requires too large a mathematical apparatus. To give at least some general characterization of world equations, let us make a few comments. World equations are differential equations whose unknown functions are, firstly, the quantities constituting the fundamental metric tensor and, secondly, g_{ik} the quantities which characterize matter. The independent variables in these equations are the world coordinates x_1, x_2, x_3, x_4."

Friedmann stresses: "The establishment of them [these equations – A. C.] is a hypothesis which, after being suitably interpreted, should be experimentally verified."

The verification of theory by experiment was even then being discussed in quite specific terms, with reference to three effects of the general theory of relativity which were later to become famous: the motion of the perihelia of planets (especially that of Mercury, as the planet closest to the Sun), the deflection of light in the field of a gravitating mass, and the gravitational red shift.

As to the perihelion motion, the question was made clear immediately. The additional perihelion motion which remained unexplained (and unexplainable) in the classical Newton theory is known, for Mercury, from numerous precise observations. "Studying the motion of a point near the Sun's gravitating mass, Einstein discovered in the first approximation Newton's law, as was expected; but there was also some correction added to the law. The application of this correction actually gave the residual motion of Mercury's perihelion, which had remained unexplained." According to the observations it constitutes 43″ (seconds of arc) in 100 years. In Einstein's calculations it was 42.89″. As Friedmann notes, "the closeness of these two numbers becomes especially remarkable if one remembers what a roundabout way brought us to the latter number."

The verification of another effect – the deflection of light in the field of a gravitating mass – was begun in 1919, when, during a solar eclipse, astronomers photographed the stars around the Sun and compared the picture with another one of the same stars in the absence of the Sun. The visible shift of the stars from their places turned out to be in good agreement with the predictions of Einstein's theory. Since that time precision of measurement, and hence reliability of verification of the theory in these observations, has been constantly increasing. Of late, such observations have been conducted outside solar eclipses, using not visible light, but radio-frequency radiation from celestial bodies. The radio-frequency radiation of quasars – the most powerful emitters in the Universe, discovered in 1963 – is quite distinguishable against the background of the Sun's radio-frequency radiation. Radio waves passing near the solar disk are deflected exactly in the way predicted by the general theory of relativity.

A remarkable example of this effect, predicted by Einstein and recently discovered, is the phenomenon of a gravitational lens. In this case a light ray from a quasar is deflected not by the Sun, but by a distant galaxy located between us and the quasar. Light can reach us in two different ways, which creates a kind of mirage for the observer: instead of one source we see two sources in the sky. Incidentally, distances to quasars are determined by astronomers using Friedmann's cosmological theory (by the velocity of their recession).

Finally, the third classical effect – the gravitational red shift of light-frequency – is observed in the spectra of some celestial bodies and, what is more important in this case, even directly in the laboratory. In the 1960s, R. V. Pound and G. A. Rebka (USA) experimentally studied changes in the frequency of light in its propagation from above downward and from below upward through 20 meters in the Earth's gravitational field. They managed to observe and measure accurately the small frequency shift, which turned out to be in good agreement with the theoretical prediction.

These are achievements of the last decades. But as early as 1922, Friedmann had reason and confidence enough to say: "Einstein's theory in its general features has brilliantly passed the test of experiment, having not only explained much of what seemed inexplicable, but having predicted, following the example of classical theories, a number of new phenomena."

With this quotation we shall conclude our short account of the book

The World as Space and Time. It is to be noted, however, that we have postponed the subject which is the most important for us – cosmology. In his book, regrettably, Friedmann gives little space to it – less than seven pages. Meanwhile, by September 5, 1922, when the date was put under the last line, work on the new cosmology was in full swing: not long before, on May 29, 1922, the first of his two cosmological articles had been finished, and the other was completed by November 1923. It remains for us only to speculate that cosmology was to become the subject of a special book or a major review article. One list of Friedmann's works has a reference to a manuscript entitled *The Universe.* We have not succeeded (only so far, hopefully) in finding this work . . .

There is one more topic that has not been dealt with in our account. In his book Friedmann discusses with great enthusiasm Weyl's theory – one of the remarkable physical theories developed right after the appearance of the general theory of relativity, in order to provide a geometric explanation of not only gravitation, but also of electromagnetism.

Weyl invented a geometry which is more general than Riemann's. His geometry allowed a new "freedom," which manifested itself during a parallel vector transfer. In a curved space, as we have mentioned before, a vector passing along a closed contour changes its direction. Weyl's theory allows a change not only in the vector's orientation, but also in its magnitude. Weyl linked such a change in a vector's magnitude to the electromagnetic field. This link is expressed through new, more general "world equations."

Friedmann noted "the grand scale of his [Weyl's – A. C.] concept, reducing everything that takes place in the physical world to the interpretation of some properties of the geometrical world." At the same time he stressed that Weyl's theory differed from Einstein's theory in that the former lacked experimental support.

Friedmann was remarkably consistent in establishing direct contacts between theory and experiment, in advocating direct experimental verification of physical concepts. This applies to all the problems on which he worked – from bombing-charts to atmospheric dynamics, and cosmological geometry. "My business is to solve equations, it is up to physicists to decide what the solution means." Friedmann's saying is only a joke, which has sometimes been taken too literally.

As for Weyl's theory, it was too much ahead of its time to be true. Indeed, the weak and strong interactions had not yet been discovered. As was found later, electromagnetism is much closer in its nature to the weak interaction than to gravitation. The unified theory of the electromagnetic and weak interactions was advanced in the mid 1960s and was later supported experimentally. It is based on the idea of the so called gauge fields, which originated in Weyl's theory. The latest advances in the physics of elementary particles are also related to the notion of gauge fields. It is in this direction that scientists are searching for a theory which would allow the nature of all four physical interactions, including gravitation, to be explained from a single point of view.

9

Geometry and dynamics of the Universe

"The surest and deepest way to study the geometry of the world and the structure of our Universe with the help of Einstein's theory consists in the application of this theory to the whole world and in the use of astronomical research." These words are from the last page of *The World as Space and Time*. Friedmann is speaking here about the cosmological problem, about the application of the general theory of relativity to the study of the world as a whole, of the world considered as a single physical system. By September 1922, when these words were being written, three different attempts had been made to solve the cosmological problem: one by Einstein himself, another by the Dutch astronomer W. de Sitter, and the third by Friedmann in the first of his two cosmological works. Cosmology was taking its very first steps, and the results so far did not look very encouraging. As Friedmann says after the above-quoted phrase, "mathematical analysis lays down its arms, faced with the difficulties of the question, and astronomical investigations do not yet give a sufficiently reliable basis for experimental studies of the Universe." And nevertheless: "these circumstances cannot be seen as more than temporary difficulties; our descendants will undoubtedly come to know the character of the Universe in which we are fated to live . . . "

However, events developed much faster than could be expected from the initial predictions. And first of all thanks to Friedmann's work. His first study, as soon became clear, was already a major step in the right direction. Cosmology was to develop rapidly even in the 1920s. As Einstein will say later, "Friedmann was the first to set out on this road."

Friedmann begins his first cosmological article by mentioning the work of Einstein and de Sitter: "In their well-known works concerned with the cosmological problem, Einstein and de Sitter," etc. We should therefore get acquainted with them here as well. It is all the more relevant, instructive and even necessary that the most recent developments in cosmology have provided a new interpretation of the first works of Einstein and de Sitter and once again – after half a century of almost complete oblivion – made them part of living science. Now we no longer look on them as something which is interesting only from the historical point of view. Of course, they are very important historically: above all, as a starting point for Friedmann. But what is probably most interesting is that not only in the history of cosmology, but in the history of the Universe itself, there may have existed an Einstein or de Sitter stage which preceded the Friedmann stage!

Symmetries of the world

Einstein's article "Cosmological considerations on the general theory of relativity," which came out in 1917 in the *Transactions of the Prussian Academy of Sciences*, was the first work which applied the modern physics of space and time to the Universe considered as a single whole. It appears that its author was quite embarrassed by what was revealed as a result of direct application of his "world equations" to the cosmological problem. What were the premises of Einstein's views of the Universe? What caused his embarrassment?

When he began to develop a cosmology based on his newly created general theory of relativity, Einstein shared certain commonly accepted views of the Universe, which had deep roots in the entire science of the modern age. First and foremost, he held the view that the structure of the world should be as simple as it can be. Thus, he thought that the Universe as a whole should be homogeneous: the distribution of matter in it is homogeneous everywhere in terms of density. We read in his article: "But if we are concerned with the structure only on a large scale, we may represent matter to ourselves as being uniformly distributed over enormous spaces . . . Thus our procedure will somewhat resemble that of the geodesists who, by means of an ellipsoid, approximate to the shape of the earth's surface, which on a small scale is extremely complicated." The space in which matter is distributed uniformly should itself be homogeneous, i.e. uniform everywhere in its geometric properties.

Homogeneity is one of the properties of space symmetry. Transition from one place to another does not change anything in overall geometry: all places (or, mathematically speaking, all the points of the world) are equivalent. Homogeneity is a symmetry with respect to possible shifts in space. Pascal (1623–62) thought of this symmetry when he said that the world is a sphere the center of which is everywhere and the circumference is nowhere. This view of the world and our place in it, which is not special in any way, goes back to the tradition of Copernicus and Bruno.

Einstein also assumed that along with the uniformity of the world there was also isotropy, i.e. the equivalence of all directions in space. This is also a geometric feature of symmetry – a symmetry with respect to the various directions in space. Both kinds of symmetry are inherent in the world as a whole and manifest themselves in its large-scale properties. (On a relatively "small" scale, e.g. on the scale of our solar system, there is evidently no isotropy, no uniformity.)

These views by themselves did not follow from either the theory of relativity or any "first principles" of physics. They were traditional views, of an intuitive and partially philosophical nature. In present-day astronomy they are fully supported by observations: the world as a whole has actually proved to be homogeneous and isotropic. But back in 1917, there was no direct factual evidence on that point, and there could be none, owing to the limitations of astronomical equipment;

Albert Einstein; late 1910s.

telescopes could evidently not reach out far enough for cosmological purposes.

Homogeneity and isotropy were taken as assumptions, as reasonable initial premises. Although Einstein in his work speaks explicitly only about homogeneity, the metric he actually uses has isotropy. (It is proved in geometry that isotropy always implies homogeneity, but not vice versa.)

Einstein proceeds from an additional assumption: the world is assumed to be spatially closed. The three-dimensional closed space is similar to a two-dimensional closed surface – the surface of a sphere. A world of this kind is of finite volume, just as the surface of a sphere is of finite area. Spaces of both kinds are homogeneous: all their points are equivalent. The latter means, in particular, that there are no boundary points and therefore, spaces are said to be closed on themselves and have no boundaries.

It is difficult to visualize a three-dimensional closed space, but the analogy with the two-dimensional surface of a sphere helps, and the mathematics here is not too complicated. Einstein suggests one simple method of calculation leading to the metric of a three-dimensional "spherical space." One should consider the Euclidean space of four dimensions x_1, x_2, x_3, x_4 (which, of course, has nothing to do with real physical four-dimensional space–time). Its metric is

$$ds^2 = dx_1^2 + dx_2^2 + dx_3^2 + dx_4^2.$$

Then a three-dimensional sphere should be embedded in this space. The equation for such a sphere is very similar to the usual equation written in a three-dimensional Euclidean space for a two-dimensional sphere, namely, all the points of the three-dimensional spherical space satisfy the condition

$$R^2 = x_1^2 + x_2^2 + x_3^2 + x_4^2.$$

Here, on the right-hand side the usual equation of a sphere is extended by the square of a fourth coordinate treated just like the other three. The equation singles out a three-dimensional spherical volume in the four-dimensional Euclidean space. This volume is a surface (or, as it is called, a hypersurface) with respect to the space in which it is embedded.

Einstein says: "The four-dimensional Euclidean space with which we started serves only for a convenient definition of our hypersurface. Only those points of the hypersurface are of interest to us which have metrical properties in agreement with those of physical space with a uniform distribution of matter."

If one defines the variable x_4 by means of the equation of the sphere, and puts it into the equation for the four-dimensional Euclidean metric, the square of the interval will no longer contain the fictitious fourth coordinate. We shall not reproduce these calculations; it should already be more or less clear from what has been said that we shall obtain in the result a new metric, having not four but three dimensions, and that this metric will describe a three-dimensional volume – the analogue of a sphere. This three-dimensional space is non-Euclidean, just like a two-dimensional sphere. Its curvature is $1/R^2$. The quantity R which for the two-dimensional sphere plays the role of radius (let us look again at the equation of the sphere), in this case determines the curvature; it is accordingly called the radius of curvature. It also serves to express the total volume of the spherical world:

$$V = 2\pi^2 R^3.$$

Thus, the spatial properties of the Universe are described. But what are its relations with time? How does it behave in time?

Antigravity

Concerning time, there was again a long-established traditional view which Einstein followed. He thought that the Universe as a whole is in an invariable state, that it is static. In such a Universe every moment is only an instant between the infinite past and the infinite future. At each particular instant the Universe as a whole is one and the same, and therefore, the instants themselves are identical and indistinguishable. These equivalent instants constitute the time of the Universe, therefore time itself can be called homogeneous. The Universe, according to Einstein, has symmetry relative to shifts in time: at whatever moment one performs physical or astronomical observations concerning the world as a whole, their results will always be identical. No matter how we change the origin of time reference in our measurements we will always obtain the same values of time intervals.

Such concepts of the Universe and its time seemed then natural and even obvious. This apparently was the generally accepted view, evolved over the centuries through contemplation of the eternal and immutable night sky. Furthermore, this view seemed to find support in some astronomical data.

Here is what Einstein writes in the article under consideration: "The most important fact that we draw from experience as to the distribution of matter is that the relative velocities of the stars are very small as compared with the velocity of light." He does not refer to any specific observational results, but in the first few pages of his article he says this about five (!) more times: he refers twice to "the fact of the small velocities of the stars;" states that "with an appropriate choice of the system of coordinates the stellar velocities are very small in comparison with that of light;" speaks about "the case in which the matter of the fixed stars [is] distributed uniformly through space;" puts forward a condition that in theory "the stellar velocities . . . [must] not exceed the velocities observed."

But it turned out that world equations (or gravitational field equations, as Einstein called them) do not allow of a static Universe. All that we read in the article shows that to Einstein this was unexpected. Yet it was necessary to find a way out. There was a choice: either to change one's views and initial premises or to change the equations. Einstein chose the second path.

"I shall conduct the reader over the road that I have myself traveled, rather a rough and winding road, because otherwise I cannot hope that he will take much interest in the result at the end of the journey." This sentence is rather unusual, not only for the present-day dry and concise style of scientific articles, but also for the more liberal language of the scientific prose of previous decades. Einstein found it possible and necessary to write in this way and no other, since after this he makes an admission which is very important for him (and apparently forced and not very pleasant for him):

"The conclusion I shall arrive at is that the field equations of gravitation which I have championed hitherto still need a slight modification . . . " This was said at the beginning of the second section of the article, and at the end of the same section he says again that "we need a generalizing modification of the field equations of gravitation."

What sort of generalization was it? Einstein, in fact, suggested a hypothesis that along with ordinary matter, all particles of which are mutually attracted, there also exists in the world an altogether unusual "medium" creating not gravity but antigravity, i.e. repulsion. It acts on ordinary matter and is capable of compensating for the mutual attraction of its particles. This being the case, the sum of the forces applied to each particle turns out to be equal to zero, and therefore, all particles may be at rest. Since all particles are at rest, the Universe as a whole is also devoid of motion – it is immovable and static.

Such is the presently accepted interpretation (using Newton's language of forces for better illustration) of what was new in Einstein's addition to the "standard" gravitation equations. He himself presents this hypothesis in his article in a somewhat formal way. He does not speak about antigravity. He introduces an additional term into the equations, which he calls the cosmological term, and shows that the equations when thus augmented allow the Universe to be static. The compensation of gravitational force is not mentioned. The equations, in fact, link the mean density of ordinary matter ρ with the density of the hypothetical antigravitating medium $\bar{\rho}$ and space curvature:

$$\rho = 2\bar{\rho} = \frac{2}{\varkappa} \frac{1}{R^2}.$$

The quantity $\varkappa$ in this relationship is expressed in a simple way in terms of the velocity of light c and the gravitational constant G:

$$\varkappa = 8\pi G/c^2.$$

As we understand it today, the above relationship is a mathematical condition compensating for the gravitation of matter by the opposing force of a hypothetical antigravitational medium. When such compensation is achieved, the sum of the forces acting on each particle of the world does indeed become zero, and everything in the world remains immutable, everything is at rest.

Einstein's world is now static. And yet such a solution of the cosmological problem did not completely satisfy the author. The concluding sentences of Einstein's article are not too buoyant. "At any rate, this view is logically consistent, and from the standpoint of the general theory of relativity lies nearest to hand; whether, from the standpoint of present astronomical knowledge, it is tenable, will not here be discussed. In order to arrive at this consistent view, we admittedly had to introduce an extension of the field equations of gravitation which is not justified by our actual knowledge of gravitation." And finally, the additional cosmological term in the equations "is necessary only for the purpose of making possible a quasi-static distribution of matter, as required by the fact of the small velocities of the stars." It is the end of the article, but by no means the end of the story.

It is unlikely that Einstein (even he!) could guess at the time what implications his hypothesis about "universal antigravity" would have in the future. We will deal with this later, in the last chapter of the book.

De Sitter's world

The first response to Einstein's cosmological investigation was a series of articles by the outstanding Dutch astronomer and theoretical physicist Willem de Sitter (1872–1934). In an article published as early as 1917 in the authoritative (then and now) British astronomical journal *Monthly Notices of the Royal Astronomical Society* he presented a new cosmological solution to Einstein's equations with the cosmological term. De Sitter discussed the cosmological problem in his private correspondence with Einstein, and the author of the theory of relativity approved of his work. Einstein wrote later about de Sitter: "Professor de Sitter was one of the most outstanding scientists in the field of astronomy. In addition, he made an important contribution to the theory of relativity. For example, by spectroscopic observation of double stars he showed that the velocity of light does not depend on the dynamic state of the light source. De Sitter made a significant contribution to solving an important cosmological problem of space structure in the theory of relativity." The latter words refer to the investigations of de Sitter which were a new step in cosmology along the path which Einstein outlined.

Like Einstein, de Sitter sees the Universe as static and immovable. He also accepts the assumption of the homogeneity and isotropy of three-dimensional space. And now it turns out that Einstein's equations allow of a new variant of a static closed Universe. If these equations contain the cosmological term (corresponding, from our modern point of view, to a hypothetical antigravitating medium, uniform in density), the three-dimensional space should be curved like a sphere, and closed on itself even in the absence of the "ordinary" matter distributed uniformly on average across the whole volume of the world. Formally, this satisfies the condition:

$$\rho = 0, \qquad \bar{p} = \frac{3}{\varkappa} \frac{1}{R^2}.$$

As above, R is here the radius of curvature of the three-dimensional space.

The curvature of space, and hence of space–time in the absence of matter, perplexed many. It seemed to contradict the initial postulates of the general theory of relativity. Its most important principle, according to which the curvature of space–time is caused by gravitating matter, appeared to be violated. An empty space–time could not exist at all.

But the point is that, as we now understand, in the presence of the cosmological term space is actually no longer empty – it is uniformly filled by the antigravitating medium. In Einstein's world picture, this medium compensates matter gravitation. In de Sitter's world picture, the antigravitating medium is the sole "space filler." Yet, as in Einstein's theory, according to de Sitter this medium is at rest. In terms of the language of forces we may say that there is antigravity in space, which is created by this hypothetical medium, but the antigravitational force so created does not act upon the medium itself. Remaining uncompensated, this force nevertheless causes no motion of the hypothetical medium: the medium is at rest and its density does not change with time. To create a force, but not to be acted upon by it, is apparently a unique property which is inherent only in this medium.

A medium of this kind is certainly unusual. It can by no means be presented as consisting of "ordinary" particles. It can be said in anticipation that this medium is a physical vacuum; we shall discuss it in more detail in the last chapter of the book, in which we shall again turn to de Sitter's world, a remarkable theoretical discovery which has now undergone a far-reaching development.

Friedmann highly esteemed the work of his Dutch colleague. Friedmann's acquaintances recall his remark, full of the spontaneity and wit so characteristic of our hero: "Your X is a runt. Now de Sitter is a real man." They say that he pronounced this name with a soft Russian "t".[1]

The new cosmology was only taking its first steps in the works of Einstein and de Sitter. Soon, however, the time of major developments in the science of the Universe was to come.

"On the curvature of space"

The long-established scientific and philosophical tradition, the accepted paradigm, described the Universe as static and invariable, which was seen as its most important inherent property. It appeared to be an extremely stable element of the generally accepted vision of the world worked out through the efforts of many generations of thinkers. This view of the world was based on ideas about the uncreatedness of the Universe; it was thought for many centuries that the scientific approach to the Universe requires recognition of its eternity and identity to itself at all times. Einstein, as we have seen, also proceeded from this assumption; his point of view was shared by de Sitter and, as far as we know, by all other physicists and astronomers who were concerned at the time with cosmological problems. In the new physics they tried to retain the traditional

[1] Another play on words: the real man was de Sitter or ten (*desyat'*) of the other fellow.

concept without subjecting it to revision or serious verification. While citing astronomical evidence to support his views, Einstein did not, in fact, have any doubts about these views; as for the data he refers to, as a matter of fact they were not directly related to cosmology.

The new physics, however, brought an essentially different conception of the world. It led to a radical break with the tradition which had been dominant in the scientific community, to a drastic change in the previous system of views. This is what Friedmann accomplished in his investigations. He discovered dynamics and development in the Universe. His works created the evolutionary cosmology which became the genuine science of the Universe.

Friedmann's two cosmological articles written in 1922–23 are devoid of loud declarations; they have quite modest titles. (It will be recalled that Copernicus' book also had an innocent title, *On the Revolutions of the Heavenly Spheres*.) They were both published in the most authoritative physics journal of the time, *Zeitschrift für Physik*. The first of them, finished by the author on May 29, 1922, was published in this journal in the same year; two years later it was published in Russian in the *Journal of the Russian Physico-Chemical Society (Physics Section)*.[2]

The article was entitled "On the curvature of space." It was not so much about curvature as about the dynamics of the world. Friedmann mentions at the beginning of the article the works of Einstein and de Sitter, his two predecessors in cosmology, and immediately formulates the exact purpose of his investigation: "indicating the possibility of obtaining a particular world, the space curvature in which, being constant in respect to three coordinates taken to be space ones, changes with time, i.e. depends on the fourth taken as a time coordinate."

The presentation of the material follows a simple and clear logic: initial assumptions, solving equations, discussion of results.

Friedmann's geometrical assumptions are the same as those of Einstein and de Sitter. Specifically, it is assumed that the three-dimensional space is similar to a sphere and has everywhere the same curvature proportional to $1/R^2$, where R is the radius of curvature. The metric of the four-dimensional space–time is given by the following expression for the squared interval:

$$ds^2 = c^2 dt^2 - dl^2,$$

where c is the velocity of light, and

---

[2] Zhurnal Russkogo Fiziko-Khimicheskogo Obshchestva (Chast Fizicheskaya), predecessor of the modern *Zhurnal Teoreticheskoi i Eksperimentalnoi Fiziki*.

$$dl^2 = R^2[\sin^2\chi(d\vartheta^2 + \sin\vartheta^2 d\varphi^2) + d\chi^2].$$

We have ventured to change the notation slightly here, so as to make it the same in this chapter as in the previous one. In Friedmann's article x_1, x_2, x_3, x_4 are used for the three space coordinates and time, instead of the symbols ϑ, φ, χ and t. This does not, of course, make any essential difference; and our notation is closer to that used in today's scientific literature. We used angular coordinates ϑ and φ when (in the previous chapter) we dealt with the metric of a two-dimensional sphere. The new coordinate χ is a third space coordinate which is added when a two-dimensional sphere is replaced by a three-dimensional one. Since all the distances in the spherical space are limited, this third coordinate, like the first two, varies within a limited range of values – from zero to 2π.

An interval (as was mentioned in the previous chapter) is the distance between two events in four-dimensional space–time; these events are separated in time by a small (infinitesimal) interval dt, and in space by a distance dl. The latter is determined – according to Pythagoras' three-dimensional theorem – by the (infinitesimal) differences of coordinates $d\vartheta$, $d\varphi$, $d\chi$; this is reflected by the above-given formula for dl. As regards the interval ds, it is constructed "almost" according to Pythagoras' theorem; the difference is that the square of the "hypothenuse" ds^2 is not the sum but the difference of the squares of the "sides" c^2dt^2 and dl^2. This general aspect (which we did not mention) reflects another essential difference between time and space. In space–time the property of invariance belongs to a certain interval in which the signs of the space and time terms are different.

The time term in the square of the interval is very simple; as regards the space term, a few words should be added. If we introduce spherical coordinates not in the spherical, but in the ordinary Euclidean three-dimensional space, then in the expression for the interval dl, where we had $\sin^2\chi$, we shall have just χ^2. And the coordinate χ will vary not within a limited range, but from zero to infinity. This underlies all the differences between the curved spherical space and Euclidean space. First of all the volume of the spherical space is finite (see above). But in the spherical space, relations between the various geometrical quantities also change. For example, the ratio of the circumference of a circle to its radius will be not 2π, as usual, but $2\pi\sin\chi/\chi$, i.e. less than 2π. Similarly, the ratio of the area of a sphere's surface to its squared radius will be not 4π, but $4\pi(\sin\chi/\chi)^2$, which is less than 4π. All this, as well as the geometrical properties of the three-dimensional space, can be derived by considering the squared distance dl^2. If one takes, for instance, two points which have the same coordinate χ and differ only in the angles ϑ and φ, one will obtain for the distance between these two points the same expression as was obtained in the previous chapter for the metric of the two-dimensional sphere. The only difference is that now $R\sin\chi$ and not R is the "radius" of the sphere. But the real radius of the sphere, i.e. the distance from the origin of coordinates to all its points, is the quantity $R\chi$, as can be seen from the general formula for dl. This brings about the above-given ratio $\sin\chi/\chi$.

Let's go back to the text of Friedmann's article and mention the other assumptions, besides geometrical ones, introduced by him. Matter is, on average, uniformly distributed in space. The pressure of matter is neglected, i.e. pressure is taken to be far less than the energy density of matter. Both these physical quantities feature in the general theory of relativity as the sources of the gravitational field, or, in other words, are the cause of space–time curvature. Although they are most often measured in different units, they have the same physical dimensionality and can indeed be compared with each other. Pressure is always negligible compared with energy density in "ordinary" states of matter, when its particles move at velocities much smaller than the velocity of light.

When the initial assumptions have been formulated, the author can specify in what way his paper will differ from those of his predecessors. Friedmann points out that Einstein's world represents a particular case, for which "R is a constant (independent of x_4!) radius of space curvature." It will be recalled that in Friedmann's theory x_4 is time. And his exclamation here, which is quite in place, is the only digression from the simple, calm and reserved style of the article. For this is the point in which Friedmann differs from Einstein. Friedmann does not impose on the curvature radius the condition of being constant in time. And the problem is to find out how the radius of space curvature, the world's volume, and also the density of the matter distributed within it, can change in time.

To solve this problem, Friedmann uses Einstein's world equations in their most general form, including the cosmological term. The latter can, in particular, be equal to zero – this is allowed by the formulation of the problem. The author reduces the ten world equations to two equations in accordance with the number of independent unknown quantities: there are two of these, the radius of the world and the density of matter. One of the equations represents the relationship between the radius and the density. And the relationship is quite natural: indeed, if the total mass of matter in the world is M, its density is obtained by dividing this mass by the volume:

$$\rho = M/V, \qquad V = 2\pi^2 R^3.$$

The second equation allows us to obtain the radius of curvature as a function of time. Its solution is written in an implicit form (because the integral obtained is non-elementary). But it does not, of course, create any difficulties for Friedmann, and he performs a clear and consistent analysis of the behavior of the curvature radius in time. We will not reproduce this analysis here and will only present its results.

It turns out that the radius of curvature can be either an increasing or a periodic function of time. An increase (monotonic, i.e. perpetual) is possible when the "cosmological term" corresponds to a positive density of the vacuum energy (in the modern terminology which we introduced in the previous chapter). The periodic behavior of the radius in time, i.e. first an increase, then a decrease, and then again an increase, etc., is possible with a negative or zero vacuum density.

If the radius of curvature always increases, the world's volume is always increasing, too. As a result of limitless expansion, the volume can become infinitely large. If the radius of curvature first increases and then decreases, the world's volume, too, first increases and then decreases. In this latter case the world's volume never exceeds a certain maximum value. Its increase and decrease – the expansion and contraction of the world – can, in principle, alternate in such a way that the world will pulsate, repeating the cycles of extension and contraction for an indefinite period of time.

As to the density of matter in the world, its behavior in time appears to be wholly dependent on the way the world's volume changes. A world of finite volume contains a finite mass of matter which (in Friedmann's solution) always remains the same. Therefore, the density should change in inverse proportion to the cube of the radius of curvature. If the radius increases perpetually, the density drops and becomes infinitesimal (remaining, of course, positive), while the world expands infinitely. In the pulsating world the density first drops, and then again increases, etc.

It may be worthwhile to interrupt the presentation of Friedmann's paper at this point, and look at what the author writes on the same subject in his popular book. As has already been said, the cosmological section of the book is very brief. Friedmann formulates the results of his first cosmological paper in just a few sentences:

One can arrive at basically two types of Universe: (1) the stationary type, in which space curvature does not change with time; and (2) the variable type, in which space curvature changes with time. The first type of Universe can be pictured as a sphere, the radius of which does not change with time; the two-dimensional surface of this sphere will be exactly the two-dimensional space of constant curvature. On the other hand, the second type of Universe can be pictured as an ever-changing sphere, now spreading itself out, now diminishing its radius and as it were compressing itself. The stationary type of Universe comprises only two cases, which were considered by Einstein and de Sitter. The variable type of Universe represents great variety of cases; there can be cases of this type when the world's radius of curvature . . . is constantly increasing in time; cases are also possible when the radius of curvature changes periodically: the Universe contracts into a point (into nothing) and then again increases its radius from a point up to a certain value, then again, diminishing its radius of curvature, transforms itself into a point, etc. This brings to mind what Hindu mythology has to say about cycles of existence, and it also becomes possible to speak about "the creation of the world from nothing," but all this should at present be considered as curious facts which cannot be reliably supported by the inadequate astronomical experimental material.

With these words (the reader will probably agree they were worth quoting fully here), Friedmann formulates his main scientific result. But he also has the courage to speak of some far-reaching implications which follow from it. Creation of the world from nothing is the most important implication and at the same time an exceptionally complex problem which Friedmann places before physical science. He speaks about it as if about a curious fact . . . But what could be said to the reader in the early 1920s about this newly emerged grandiose problem, if it is still far from being solved now, 65 years later? One can consider it a significant achievement of the most recent years that this problem is at last being given a rigorous physical formulation in terms of quantum theory and the theory of relativity, and is becoming the subject of specific theoretical studies. This reveals a profound inner relationship between the physics of the largest object of science – the Universe as a whole – and the physics of the minutest bodies in Nature – elementary particles. It is no exaggeration to say that the future of fundamental physics will directly depend on the way it tackles the problem of the early Universe as posed by Friedmann. We shall come back to this issue in the last chapter of the book and will only note here that it is increasingly recognized now as the major problem not only of physical science, but of the whole scientific world view.

The Universe in time

The expansion of the world begins from the state where, in Friedmann's words, "space was a point ($R = 0$)." This means, in particular, that the initial density of the world was infinitely great:

$$\rho \propto R^{-3} \to \infty \text{ as } R \to 0.$$

Such an unusual, entirely unique state of the world and matter is called a cosmological singularity. The discovery of the initial singularity is one of the most remarkable achievements of Friedmann's theory. Friedmann does not dwell too much on this question in his article (nor in the book either), only stating the fact as such; in the last chapter of our book we shall discuss in brief the present-day understanding of the problem of cosmological singularity.

How far away from us in the past was the moment of the beginning of expansion? "Using an obvious analogy," writes Friedmann, "we shall refer to the time it takes the radius of curvature to increase from zero to R_0, as the time which has passed since the creation of the world." R_0 here is the value of the radius of curvature at the present time. Friedmann gives a general formula to calculate the time interval separating us from the beginning of the world's expansion; as to the numerical estimate resulting from it, he writes: "The data available to us are completely inadequate for any kind of numerical estimates and for solving the problem of what kind of world our Universe is."

And yet the characteristic cosmological time appears in the paper. Warning that any figures can at present be of only illustrative, tentative value, Friedmann gives his estimate of a possible period of expansion–contraction for the case of a pulsating world. Neglecting the cosmological term and considering the total mass of the world to be "the mass of 5×10^{21} of our Suns, we shall obtain for the age of the world a quantity of the order of ten billion years."

Thus there emerged, for the first time in science a quantitative characteristic of the behavior of the Universe in time, a numerical measure of its dynamics.

Friedmann obtains this characteristic time using the formula for the period of the pulsating world, according to which its value is expressed in terms of the world's total mass:

$$T = \frac{4}{3}\frac{GM}{c^3}.$$

(We have written this formula in a slightly changed form; c here is the velocity of light in a vacuum). As we have said, this relationship was obtained by the author neglecting the cosmological term. Therefore, to estimate the period one needs to know only one quantity – the world's total mass. But where does the value of 5×10^{21} solar masses for the mass M come from?

Friedmann does not refer to any astronomical data, and these are indeed "completely inadequate for any kind of numerical estimates." However, the question of the total mass had been raised in scientific (and even more often in popular science) literature. The world's total mass, but without a numerical estimate, is also dealt with in Einstein's first cosmological work. As is always the case when there is wide interest in a problem, a kind of "science folklore" appears: people talk (but less often write) about some probable figures, about more or less plausible values; there appears an opinion shared by many as to what can be considered acceptable and what is ruled out completely, etc.

Some figures lacking in the article are given by Friedmann in his book. He writes that Einstein "determined, in accordance with astronomical data, the radius of curvature of the Universe to be 10^{12}–10^{13} times the distance from the Earth to the Sun (let us remind the reader that the mean distance from the Earth to the Sun is 1.5×10^{13} cm), and the density (everywhere constant) to be 10^{-26} g/cm^3." There is no exact reference to Einstein's work; there are no such data in the article cited by Friedmann in the work under consideration.

Einstein generally avoided such "details" at the time, believing that a quantitative concretization of the theory was a matter for the distant future, and did not seem to be too keen even on that. Great interest in astronomical data was displayed by de Sitter. His investigations mention estimates of the maximum distances accessible to astronomical observation, and attempts to determine the real density of matter in the world. It cannot be ruled out that it was from de Sitter (and not from Einstein at all) that Friedmann borrowed the astronomical data he used for his estimates. It is also probable that he simply took them from scientific "folklore," and "folklore" gives all or almost all the credit to its most popular hero.

One can judge the state of the art in observational cosmology in the early 1920s by Arthur S. Eddington's remarkable book *The Mathematical Theory of Relativity*, whose first edition came out in 1923. The second edition, published in 1924, was translated into Russian and published in this language in 1934 under the title *Theory of Relativity*. In his presentation of the cosmological problem, Eddington tries to introduce into the discussion all the astronomical material available. He says that "our knowledge of the stellar Universe extends nearly to 10^{25} cm" and the latter value is taken as a tentative estimate of the world's radius. This quantity is 10^{12} times the distance from the Earth to the Sun. As far as the average density of matter in the Universe is concerned, the estimate was derived from data on the density of matter "near the Sun," i.e. the density in our Galaxy, approximately 10^{-24} g/cm^3. The average density of matter within the Galaxy is still estimated at this value today. Allowing for the possible existence of areas with less matter than near the Sun, a value of approximately 10^{-26} g/cm^3 was accepted for the world as a whole. This estimate seems to have held till the beginning of the 1930s (the present-day estimate is 5×10^{-30} g/cm^3). In his popular lecture "The present state of the theory of relativity," delivered in October 1931, Einstein says that "the average density is represented by a fraction whose numerator is 1, and whose denominator is 1 with 26 or 27 zeros."

If one assumes the world's radius to be about 10^{25}–10^{26} cm and its density to be 10^{-26} g/cm^3, then using the formula linking these quantities with the total mass $M = 2\pi^2 \rho R^3$ one will obtain $M = 3 \times 10^{17}$ to 3×10^{20} solar masses, which is at least an order of magnitude less than 5×10^{21} –

Friedmann in 1922 or 1923.

the figure used in Friedmann's article. With a lesser mass the period of the pulsating world should be correspondingly less. But Friedmann evidently preferred the value of ten billion years. It appears in his popular science book too, where he uses not the period of pulsations, but another measure of the dynamics of the world and a more interesting one at that – the time that has passed since the Universe began to expand.

It is useless, in view of the lack of appropriate astronomical data, to cite any figures characterizing the "life" of the changing Universe; if one nevertheless begins to calculate, out of curiosity, the time which has passed since the moment when the Universe was created out of a point up to its present state, begins therefore to determine the time which has passed since the creation of the world, one will obtain figures in tens of billions of our ordinary years.

The greatest time periods which were the subject of discussion in 1920s concerned the age of the Earth, the Sun and the stars. They could serve as a kind of reference point for cosmologists, who might reasonably believe that the age of the world is at any rate no less than the age of astronomical bodies. In the above-mentioned 1931 lecture Einstein cites the age of the Earth as the greatest of the independently measured time periods: he says that the age of the Earth, determined using the radium method, is about eight hundred million years. Present-day estimates of the age of the Earth give a figure of about 4.5 billion years. Eddington in those years had another figure which was greater and fairly accurate, and is accepted now: he spoke about the age of the Sun as being five billion years in his 1930 article (which was published as a supplement to the Russian edition of his book).

It is most likely that Friedmann did not have data about the age of celestial bodies, otherwise he would probably have mentioned them, at least in his popular science book. But he was definitely sympathetic toward the figure that he used twice – around ten billion years. It was, however, no more than a conjecture. A happy conjecture, let us add, which was later justified. Present-day estimates of the age of the world, based on data on the prevalence of radioactive nuclei and the evolutionary state of certain star clusters, give a value of 17 to 20 billion years.

"On the possibility of a world with constant negative curvature"

This was the title of Friedmann's second cosmological paper, which he finished in November 1923. It came out in January–February 1924 in the same journal as his first work. The second work deepens and develops his ideas about the dynamics of the world and its geometry.

It is to be noted that judging by the titles which the author gave to his two articles, his approach was primarily based on geometry. In the second work he takes a further step: he breaks away from the geometrical constraint adopted by Einstein and de Sitter. Previously, only worlds with a three-dimensional space similar to a sphere had been considered. Friedmann introduces into cosmology a new three-dimensional space which also has homogeneity and isotropy, but is of another geometrical type, namely of that of Lobachevsky's geometry.

Let us recall how in geometry one distinguishes spaces by type of curvature and let us do so, for simplicity, as above, using two-dimensional spaces, i.e. surfaces. Let us imagine that a plane touches a spherical surface at some point. It is clear that the whole sphere lies entirely on one side of such a plane. Let us now take a surface having the shape of a mountain pass (Fig. 2). Let us imagine a plane which touches this surface at the very point of the pass, i.e. where the height of the mountain ridge is least. Such a plane lies horizontally. But it must now dissect the surface, and different parts of the surface will find themselves on different sides of the plane.

These two cases are commonly distinguished, the curvature of the surface being ascribed different signs. The curvature of wholly convex surfaces is considered to be positive, while the curvature of saddle-shaped surfaces is negative. A sphere is an example of a two-dimensional space with everywhere uniform positive curvature. An example of a two-dimensional space with everywhere uniform negative curvature is a one-sheet hyperboloid, which looks exactly like a mountain ridge with a pass; another example is a pseudosphere having the shape of a gramophone horn. This second type of geometry was discovered by Lobachevsky.

Having studied the world of a three-dimensional space of positive curvature, Friedmann extends the concept of dynamics to the world with a three-dimensional space of negative curvature. This work seems to have been prompted by discussions with his colleagues; the author writes in a footnote: "the possibility of a world with negative curvature of space requires a separate study, and this was pointed out to me by my friend Professor Ya. D. Tamarkin." One particular variety of a static world with negative curvature of space, which is mentioned in the work, was studied at the suggestion of Friedmann's young colleague: "The possibility of this case was pointed out to me by V. Fock." Academician Vladimir Alexandrovich Fock wrote in 1963: "I remember these works well, especially as the second of them (and perhaps both of them) was translated by me into German for A. A. Friedmann, and I also drew his attention to one case

Fig. 2   Saddle-shaped surface: a two-dimensional space of
negative curvature.

of negative curvature space he had not considered." (In Friedmann's
Collected Works, the second article is in reverse translation from the
German, edited by V. A. Fock.)

Friedmann shows that static solutions for a world with negative space
curvature are obtained only with matter having zero or negative density;
the latter possibility has no physical meaning, while the former corresponds to a non-zero cosmological term, i.e. to vacuum as it is understood
today. (As was subsequently shown, such a static world is a part of de
Sitter's world if the vacuum density is positive.)

The metric in which a three-dimensional space has an everywhere
uniform negative curvature is written by Friedmann as

$$ds^2 = -\frac{R^2}{c^2}\frac{dx_1^2 + dx_2^2 + dx_3^2}{x_3^2} + dx_4^2.$$

As in the first work, the fourth coordinate plays the role of time,
determining the radius of curvature R. As to the spatial part of the metric,
if written in the "spherical" coordinates used in the first article, it would
take the same form as for the spherical space – with a single, but
significant, difference: where in the spherical metric there was a trigonometric sine of a radial coordinate, there should now be a hyperbolic
sine. And the radial coordinate itself can now vary within the range from
zero to infinity. The replacement of $\sin\chi$ by $\mathrm{sh}\chi$ in the interval brings
about a change in such geometrical relationships as the ratio of a circle's
circumference to its radius and the ratio of a sphere's surface area to the
square of its radius. For a space with negative curvature these ratios are
not less but greater than 2π and 4π respectively.

The type of spatial metric which was used by Friedmann was borrowed
by him from a book by the Italian mathematician L. Bianchi published in
Bologna in 1923 – Friedmann makes a reference to this book, the only one
in his paper, except mentioning his first paper. (It will be recalled that
Friedmann studied Bianchi's books when at the Gymnasium.) There
were, incidentally, more references in his first paper – to works by Ein-

stein and de Sitter, to Eddington's 1921 book *Space, Time and Gravitation* (published in Russian in Odessa in 1923), and to the German physicist M. Laue's book *Theory of Relativity*, published in German in 1921; the German mathematicians K. Weierstrass and J. Horn were mentioned as well.

The dependence of the radius of curvature R on time is determined by Einstein's equations, and Friedmann gives the corresponding solution (without analyzing it, though, or studying the difference from the case of positive curvature). He also finds the dependence of matter density upon the radius of curvature

$$\rho \approx A/R^3,$$

which reminds one of the corresponding formula for the world with positive space curvature, the difference, though, being that A is now an arbitrary constant, while previously it was expressed in terms of the total mass contained in the spherical space (but the author does not find it worth dwelling upon this either).

Summing up, Friedmann stresses that the density in his solution is positive, as it should be: "From this follows the possibility of non-stationary worlds with constant negative space curvature and positive matter density." ("Constant" means here independent of space coordinates, i.e. identical, for a given moment of time, at all points.)

One can see that Friedmann was primarily concerned in this work not with the dynamics of the world – this had already been discovered by him; he was interested in the question of the finiteness or infiniteness of space. If one follows Einstein, the three-dimensional space of positive curvature should be regarded as similar in all respects to a sphere: not only the metric, but the general structure of the space as a whole; its topology should be the same as that of the sphere. In particular, the total volume of the space should be finite, just as the surface of the sphere is finite. Such are the geometric models of the world considered by Einstein, de Sitter and also Friedmann himself in his first work. In his new work he introduces a new geometry, its two-dimensional analogs no longer being similar to a sphere in terms of their general structure: paraboloid as well as pseudosphere are surfaces which are infinitely extendable, so their area should be considered as infinite. If one transfers this property of two-dimensional spaces of negative curvature directly to a three-dimensional space of the same type, one should ascribe an infinite value to its volume.

At any rate, the geometrical argument which Einstein had for asserting the finiteness of space now disappears: the new geometrical picture disproves finiteness of volume for spaces with negative constant curvature. However, it is not only a matter of metric, which may be similar or dissimilar to the metric of a sphere. The general structure of space, its global properties, its topology actually require a separate approach. Simple geometrical analogs are not sufficient here, and Friedmann finds it necessary to insist on that with some firmness both in his article and in his popular science book.

At the very beginning of his book, having mentioned cosmology for the first time, Friedmann says: "I consider it particularly necessary to treat this question about the Universe in due detail, because ... distorted ideas have spread about the finiteness, closedness, curvature and other properties of our space, which are supposedly established by the relativity principle." And on the last pages of the book we read about "one misunderstanding repeated not only in popular science articles and books, but also in more serious and specialist works on the relativity principle. I mean the notorious question of the finiteness of the Universe, i.e. of the finiteness of our physical space filled with shining stars. It is claimed that having found a constant positive curvature of the Universe one can conclude that it is finite, and above all that a straight line in the Universe has a "finite length," that the volume of the Universe is also finite, etc. This statement can be based either on a misunderstanding or on additional hypotheses. These by no means follow from the metric of the world, but the metric can only be derived from the world equations. Simple examples can convince us of that. The metric of the surface of a cylinder and the metric of a plane are identical; at the same time, on a cylinder there are "straight lines" of finite length (a circle, see Fig. 22),[3] while there are no such straight lines in a plane."

It is a remarkable example. It is easy to see that the metrics of a cylinder and a plane are the same; the cylinder can be "glued" out of a plane. A plane has a Euclidean metric. If one draws a triangle in a plane, on a sheet of paper, nothing special will happen to the triangle if the sheet edges are glued together. The sum of the angles of the triangle will remain equal to two right angles. Pythagoras' theorem, which holds for a Euclidean plane, does not lose its validity if a cylinder is made of this plane. It is this circumstance which is reflected by the metric which, as we have said many times, represents Pythagoras' theorem as an expression for the square of the distance between two points which are very close to each other.

[3] Fig. 3 reproduces that figure from the book *The World as Space and Time*.

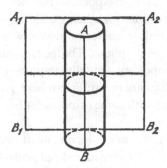

Fig. 3   A cylinder made from a flat rectangle; when the cylinder
is assembled, the lines $A_1 B_1$ and $A_2 B_2$ are made identical:
$$A_1 B_1 = A_2 B_2 = AB.$$

But the metric describes only the "local" properties of a surface (or
space), i.e. the relationships between points which are very close indeed.
As regards points remote from each other, the metric alone is not sufficient to describe these relationships. One should know, as in the example
given by Friedmann, the general design of the surface. All straight lines
on a plane can be continued ad infinitum – but only if the plane itself
extends infinitely in all directions. If a cylinder is made from a plane,
straight lines going perpendicular to its axis will become closed on
themselves. In this direction the two-dimensional space – the surface of
the cylinder – is finite and closed.

Using the example with a plane and a cylinder, Friedmann concludes:
"Thus, the world's metric alone does not enable us to solve the problem of
the finiteness of the Universe. To solve it, we need additional theoretical
and experimental investigations."

Here is one more instructive example which Friedmann cites not in the
book, but in the second article: "We call a space finite, if the distance
between two non-coincident points does not exceed a certain positive
constant number, whatever this pair of points may be. Therefore, before
considering the problem of the finiteness of space, we should also agree
what points of this space should be considered different. For example, if
we consider the surface of a sphere in a three-dimensional Euclidean
space, points lying on the same latitude and having a longitude difference
of 360° will be considered as coincident; if on the contrary we regarded
these points as different, we should get a multi-sheet spherical surface in
Euclidean space." The distance between two arbitrary points on a sphere
is always less than a certain number – the length of a great circle. But this

is only true when, as is usually the case, a sphere is not considered as consisting of many sheets. If one assumes such a multi-sheet structure and, contrary to the usual practice, considers points with longitude difference of 360° lying on the same parallel as non-coincident, "the distance can be made infinitely large (attributing the points to different sheets in an appropriate way). It is therefore clear that before we start to discuss the finiteness of the world, it is necessary to specify what points should be considered as coincident and what points as different."

Going back to Friedmann's book let us find in it a thought experiment which would make it possible to verify directly whether a spherical space is multi-sheet or "ordinary." It is suggested that we imagine "we have become unnaturally flat and are living on a large sphere, like surface shadows."[4] Having posed the question of finiteness of the space on (or in) which they are living, the shadows "could solve this problem by sending one of their spherical travelers off on a journey. Always keeping to a straight line and walking along it in the same direction, our traveler would observe the character of the landscape around him and would see that this character would change continuously during his journey; he would come across other landscapes and spherical cities which would remind him little of the cities of his native land; but approaching his home town, from which he had started, from the other direction, the traveler would notice that the surrounding terrain was becoming more and more similar to the one which he left when he set out on his long journey. On returning to the starting point, the traveler would find, through observations, that the point which he reached coincided completely with the point from which he had started; in this way . . . the finiteness of the sphere of the universe would be proved." This would be the situation in the "ordinary" case. But in the case of a multi-sheet sphere our imaginary traveler would walk continuously, receding further and further away from his "native place." Having covered a distance longer than the length of a great circle without returning home, he would understand that his space had a more complicated structure than the "ordinary" sphere. And he would no longer have any grounds to believe that the space in which he lived was finite.

Finiteness and infiniteness is another point of Friedmann's dispute with Einstein. The world is not static, as was proved in the first article. And it is not closed either – that was proved in the second one. Speaking about

---

[4] At this point Friedmann gives the following foot-note: "This comparison was given by Prof. O. D. Khvolson in one of his brilliant popular lectures on the relativity principle." In actual fact, these "shadows" came from Einstein's first popular science book and have been roaming in the literature ever since.

closedness of space in his model of the world, Einstein makes implicit assumptions about the general structure of space. These assumptions by themselves do not follow from the general theory of relativity and, in fact, they cannot follow from it, because this theory is capable of studying only local, not global, properties of the world.

In his article, Friedmann poses the question of the global topology of the Universe, i.e. of those of its properties, which concern not local geometrical relationships in small areas, but general properties of all space as a whole. This question lies beyond the general theory of relativity. It falls within the scope of a science which would describe the global topology of the world. Like general relativity it could relate the topology of space (to be more exact, of space–time) to global physical processes embracing the world as a whole. And Friedmann was the first to state clearly and definitely the basic limitations of the cosmological theory based only on general relativity. Solving the "eternal" question of the finiteness or infiniteness of the Universe is beyond its power: "Einstein's cosmological equations alone, without additional assumptions, are insufficient for reaching a conclusion about the finiteness of our world."

Who knows what ideas and projects could have originated in Friedmann's mind in connection with this problem? Could he have had a desire to overcome the limitations of general relativity, which he so clearly stated, and to construct a new theory capable of solving the question of the global structure of the world? He speaks about this subject with interest, enthusiasm, even passion – it is unlikely that he would have left it without further development, had he been granted more years to live.

Today, 67 years later, there is little we can add to what Friedmann had to say. Cosmological investigations in this field have not yet advanced very far. In the mathematical field, a very comprehensive study has been made of the relationship between the sign of the curvature of a homogeneous space and its possible topological structure. It turns out, for example, that a closed space with a Euclidean metric, i.e. of zero curvature, is possible. All in all, 18 different varieties of topology are possible for Euclidean space. For curved spaces, the number of allowable topological realizations is actually infinite.

One should note, finally, direct astronomical observations aimed at finding topological effects. Among such effects are "ghosts," i.e. multiple images of the same object in the sky. They may appear in a topologically closed space of finite volume, because the light from the radiating object may reach us in various ways; it can, for example, come from the opposite direction, passing around the world "from the other side." In this case we shall see the same object in two diametrically opposite directions. Besides searching for "ghosts," of great importance are "measurements" of the Universe in various directions. The observational program in this field was started in the Northern Caucasus by a group of astronomers headed by Viktorin Fyodorovich Shvartzman, using the 6-meter telescope (the world's largest) of the USSR Academy of Sciences. It is hoped that sooner or later definite astronomical data will be obtained concerning this question.

Discussion with Einstein

Friedmann's cosmological investigations were immediately noticed. The first of them triggered Einstein's instantaneous and very negative response: the world cannot expand, it is static. This was followed by quite dramatic events:

(1) Friedmann's first article was received by the editors of the journal *Zeitschrift für Physik* on June 29, 1922.
(2) Einstein's first response to Friedmann's article was received by the journal on September 18, 1922.
(3) Einstein's second comment about the same article was received by the editors on May 21, 1923.

The intellectual battle of the two scientists, the drama of ideas, was played in three published acts; the unity of time failed to be observed – the dispute lasted for about eight months – but the unity of place was there: the German journal.

This drama had also its back-stage part.

The *Zeitschrift für Physik* was the most popular physics journal of the 1920s. Einstein undoubtedly looked through it immediately upon receiving it. Regrettably, there is no detailed "chronicle" of Einstein's life – his "what? where? and when?" In the spring of 1922, he spent some time in France, in October of the same year he was already in Switzerland. He is likely to have had some rest somewhere (from July to December), and therefore, did not respond immediately to the article by Friedmann, of whom he is not likely to have heard before. But having read it, he seems to have decided to respond to it before leaving for the vacation. Einstein's reply was: "The results concerning the non-stationary world, contained in [Friedmann's] work, appear to me suspicious. In reality it turns out that the solution given in it does not satisfy the field equations." As Einstein says, it contradicts one of the implications of these equations, which requires the divergence of the energy–momentum tensor of matter to vanish. Einstein goes on to write that Friedmann's assumptions imply, in fact, that the radius of the world is constant in time. "Therefore," Einstein concludes, "the significance of the work is precisely that it proves this constancy."

It should be recognized that Einstein's response is somewhat disparaging in tone, and the concluding sentence annuls Friedmann's result and converts the main statement of his article into something directly opposite. Such assessment by the person who had an unprecedented authority could not but hurt Friedmann. It is unlikely that he remembered at that time that Einstein had made mistakes several times and had had to publish corrections to his papers (among his works published before 1922 and included in the four-volume collection of Einstein's selected works published in the USSR, there are six notes where he indicates his mistakes and inaccuracies).

The issue of the journal with Einstein's reply appeared in Russia in October and Friedmann and his colleagues were able to see it immediately, but this acquaintance in fact took place earlier. On December 6, Friedmann sends a detailed formal letter to Einstein's Berlin address. Here is what he writes in it (the original is in German):

Dear Professor,
From the letter of a friend of mine who is now abroad I had the honor to learn that you had submitted a short note to be printed in the 11th volume of the *Zeitschrift für Physik*, where it is stated that if one accepts the assumptions made in my article "On the curvature of space," it will follow from the world equations derived by you that the radius of curvature of the world is a quantity independent of time.

Let us interrupt the quotation. There is no doubt that "a friend of mine" is Yu. A. Krutkov. There is a note in one of his diaries: in 1922–23 he spent "a year and a day" in Germany, and the date of his departure for Russia is exactly known: September 28, 1923. Everything is thus in agreement: Krutkov came to Berlin on September 27, 1922, soon after the journal had received Einstein's comment.

In his letter Friedmann further shows by direct calculations that the necessary condition of disappearance (turning into zero) of the divergence of the energy–momentum tensor, which was pointed out by Einstein, by no means implies that the radius of the curvature of the world is constant in time. "Considering that the possible existence of a non-stationary world has a certain interest, I will allow myself to present to you here the calculations I have made . . . for verification and critical assessment."

After presenting his calculations Friedmann notes that he has recently studied "the case of a world with a constant and changing (in time) negative curvature . . . The results of the calculations have shown that . . . both the world with a constant (but negative) curvature and the world with a changing (in time) curvature may exist. The possibility of obtaining from your world equations a world with a constant negative curvature is of exceptional interest to me, and I therefore ask you to reply to my letter, although I well know that you are very busy." And he concludes: "Should you find the calculations presented in my letter correct, please be so kind as to inform the editors of the *Zeitschrift für Physik* about it; perhaps in this case you will publish a correction to your statement or provide an opportunity for a portion of this letter to be printed."

Friedmann had undoubtedly discussed his calculations with his colleagues, especially Frederiks, and was, as a matter of fact, convinced in their correctness. Let us note, however, that he did not send the letter to the editorial office of the journal, thus displaying great tact toward his opponent.

Friedmann was right in suggesting that Einstein was "very busy," though he probably hoped to receive his reply soon. However, weeks passed, and there was no answer. The explanation is very simple: Einstein was away from Berlin in December. Back in late September, about two weeks after he had sent the journal his first comment on Friedmann's article, Einstein and his wife left Berlin – first for Switzerland, and from there for France – and on October 11th, he sailed off to Japan. Einstein is known to have learnt with delay that in his absence he had been awarded the Nobel prize for physics. He could not be presented in Stockholm at the presentation ceremony on December 11. Only in March 1923, having visited Palestine, France and Spain on the way back home, did Einstein return to Berlin. It obviously took him some time to sort out the mail accumulated during his almost half-year absence . . .

April passed, and in May he was invited to Leiden (he was a Professor Emeritus of Leiden University) to attend the farewell lecture of Lorentz, who was about to retire. At the same time Krutkov was in Leiden, too – Einstein met with him at Ehrenfest's place, where he always stayed when in the Netherlands. Krutkov, whom we first mentioned in the section "Friedmann and Ehrenfest" (Chapter 3), had by 1923 become one of Russia's best-educated theorists. He was well versed in the theory of relativity from his work at the university seminar. Krutkov's acquaintance with Einstein, and his conversations with the great physicist, are reflected in Krutkov's surviving notes and his letters to his sister Tatiana

Alexandrovna Krutkova in Petrograd (his notebooks of May 1923 are covered with formulae from Friedmann's article and relevant calculations). Krutkov writes to his sister: "On Monday, May 7, 1923, I was reading, together with Einstein, Friedmann's article in the *Zeitschrift für Physik.*" May 18: "At 5 o'clock Einstein reported on his latest paper to Ehrenfest, Droste and a Belgian . . . I defeated Einstein in the argument about Friedmann. Petrograd's honor is saved!"

We have seen that Einstein's second note concerning Friedmann's cosmological article was received by the journal's editorial office on May 21, 1923. There is no doubt that this coincidence in time is not accidental and that the note was the outcome of his discussions with Krutkov.

Let us now quote Einstein's second note. It was published in Russian in his *Collection of Scientific Works* and included as a supplement in the volume of Friedmann's *Selected Works.* It reads as follows:

In my previous note I criticized the above-mentioned work [Einstein included the title of Friedmann's paper in the title of his note – V. F.]. However, my criticism, as I became convinced by Friedmann's letter communicated to me by Mr. Krutkov, was based on an error in calculation. I consider that Mr. Friedmann's results are correct and shed new light. It has turned out that the field equations allow not only static but also dynamic (i.e. variable with respect to time) centro-symmetrical solutions for the space structure.

The implication of the phrase "by Friedmann's letter communicated to me by Mr. Krutkov" was obvious: amidst the hurly-burly of his arrival from the long journey Einstein failed to read Friedmann's letter; only after his meeting with Krutkov was he aware of its contents, probably because they were conveyed to Krutkov by Friedmann. The point is that on May 7, Einstein and Krutkov were reading Friedmann's *article* – Krutkov did not say a word about the *letter* from Petrograd to Berlin.

Yu. B. Tatarinov, a Soviet historian of science, who recently investigated the dispute between Friedmann and Einstein in the light of the above-cited documents of Krutkov's, published in 1970 and 1974, decided to check the accuracy of the translation of Einstein's notes cited from the above-mentioned academic publications. The sentence in question, in a more precise translation, reads as follows: "My objection was, however – as I became convinced thanks to Mr. Krutkov with the help (*an Hand*) of a letter from Mr. Friedmann – based on an error in calculation." From this, Tatarinov draws the conclusion that Einstein did read Friedmann's letter. This was probably true. Krutkov who, according to V. A. Fock's reminiscences, was acting at Friedmann's request, met with Einstein, but in Leiden, not Berlin; they analyzed Friedmann's article and Krutkov supplemented it with Friedmann's arguments, which he knew. On May 16, 1923, Einstein left Leiden for Berlin. On his arrival he presumably began sorting out the accumulated correspondence and found Friedmann's letter (alternatively, he could have found it earlier and taken it to Leiden with him, prompted by Krutkov), and on May 21, as appears from the second note, submitted it to the *Zeitschrift für Physik*.

A recently published article by J. Stachel, dealing with the principles of selecting materials from Einstein's archive, to be published in a multivolume collection of his works (it is already being published), adds another, quite remarkable, touch to the history of the discussion between Einstein and Friedmann. In Einstein's papers there is a typewritten draft of the letter to the editors of the *Zeitschrift für Physik*. It is identical with the text published in the journal apart from one sentence, which ends the typewritten text but did not appear in the journal, having been deleted by Einstein.

Even after recognizing the absolute correctness of Friedmann's mathematical calculations and conclusions, Einstein till the very last moment was doubting their being *physically* correct and fruitful. And one might guess that at the very last moment, immediately before sending the letter to the editors, he crossed out the above-mentioned sentence. One can assume that once again he was influenced by Krutkov, with whom he was probably returning from Leiden to Berlin (this joint trip was mentioned by Krutkov in a letter to his sister on May 15). As has been mentioned above, Einstein happened to make mistakes in his publications, but, according to his contemporaries, he considered his error in assessing Friedmann's work to be most unfortunate. Krutkov seems to have helped his great colleague to avoid another mistake.

On June 1, Yuri Krutkov, now in Berlin, wrote to his sister: "Einstein is very nice . . . I am going to visit Einstein one of these days." Finally, on June 8: "How is Friedmann getting along? I have written him about his business with Einstein and keep wondering what he is thinking about. At any rate he should have written me." Regrettably, we could not find any correspondence between Krutkov and Friedmann, but other letters – from Krutkov to his sister and, particularly, from Friedmann to N. Ye. Malinina – can provide important evidence about Friedmann's stay in Berlin in August–September of that year (1923) and his attempts to meet with Einstein personally. On August 9, Krutkov writes: "Friedmann is here; today, in a few hours' time, he is going to Hamburg. Einstein's note, in which he is rehabilitated – with my encouragement – has already come out." And here is what Friedmann writes on August 19: "My trip is not going well – Einstein, for instance, has left Berlin on vacation, and I will not be able to see him"; on September 2: "I have only the following things left to be done: (1) to visit Göttingen, (2) to see Pahlen (an astronomer, my former assistant), (3) to see Mises (editor of the *Zeitschr. f. Angew. Mathem.*) and Einstein, and (4) to make necessary purchases." On September 13: "Today I visited the astronomer Pahlen, an old friend of mine, I met there the astronomer Freundlich, a very interesting person, we talked with him about the structure of the Universe . . . Everybody was much impressed by my struggle with Einstein and my eventual victory, it is pleasant for me because of my papers, I shall be able to get them published more easily."

The old friend Pahlen was Alexander Friedmann's assistant in Kiev in 1916 (see Chapter 5). Until World War I he was a Doctor of Philosophy of Göttingen University, was a member of the Russian Physico-Chemical Society, and lived in St. Petersburg. After the revolution of 1917 Pahlen seems to have returned to Germany and begun working there at the Potsdam Observatory. Additional information about him can be found in the biographical handbook *Astronomers* (Kiev, 1986) on p. 470: "Pahlen, Emanuel von der (1882–1952) – a German astronomer, worked at the Potsdam Observatory. Studied stellar statistics." In recognition of his contribution to science the name of Pahlen was given to a lunar crater. Pahlen gave much attention to the problems of cosmology and the theory of relativity, and in 1924 he published a book *The Infinity of Space and the Theory of Relativity*, which came out in Potsdam (naturally, in German). Judging by his card catalog, Friedmann had this book in his library.

Erwin Fritz Freundlich (1885–1964) was a German astronomer who was concerned with astronomical aspects of the theory of relativity; he was the founder and director of the Einstein Institute in Potsdam (as the same handbook says). In 1916 he published in Berlin a popular-science book *The Foundations of Einstein's Theory of Gravitation*, for which the author of the theory wrote a highly complimentary foreword (in 1924 this book was published in the Soviet Union in Russian translation). On the Moon there is a crater called Freundlich.

Incidentally, on the Moon there is also a crater named for Alexander Friedmann!

Friedmann did meet Mises, an Austrian mathematician and student of mechanics who concerned himself with aerodynamics (and who was also a pilot!), but the meeting with Einstein never materialized, either during this visit to Germany, or during the next one in 1924 . . .

Speaking about Friedmann's investigations in relativistic cosmology,

Professor Lev Gerasimovich Loitsyansky, who was Alexander Friedmann's close friend in the 1920s, recalls: "These investigations gave him more pleasure than any others." Witty and with no feeling of self-importance, Friedmann joked that he had "shod" Einstein – i.e. taught him the rudiments.

Legends were told about Alexander Friedmann during his lifetime. This is what Academician Isaak Kikoin wrote, recalling the 1920s when he studied at the Faculty of Physics and Mechanics of the Polytechnical Institute: "We students used to follow the scientific achievements of our professors very closely, were fans of theirs. We were very proud to have Alexander Friedmann – one of the world's greatest students of mechanics – teaching at our faculty. We heard that he had made a very significant contribution to the general theory of relativity, having corrected Einstein himself . . . We took great pride in this."

Georgy Abramovich Grinberg, a Corresponding Member of the USSR Academy of Sciences, once observed: "It was said that he himself did not realize what he had done. Nonsense! He understood quite well what he was."

Ekaterina Petrovna Dorofeyeva-Friedmann writes in her memoirs: "Always ready to learn from everybody who knew more than he did, he realized that in his work he was blazing new trials, difficult and unexplored by anyone, and he liked to quote these words of Dante's: 'L'acqua ch'io prendo giammai non si corse' (*Paradiso*, II, 7). 'The sea I sail was never crossed before'."

In the pages of his research, in the reminiscences of his contemporaries, Friedmann is seen as a profound, independent-minded, and daring thinker who destroys scientific prejudices, myths and dogmas; his intellect sees what others do not see, and will not see what others believe to be obvious but for which there are no grounds in reality. He rejects the centuries-old tradition which chose, prior to any experience, to consider the Universe eternal and eternally immutable. He accomplishes a genuine revolution in science. As Copernicus made the Earth go round the Sun, so Friedmann made the Universe expand.

10

Petrograd, 1920–24

Arrival and first months

According to various documents and reminiscences about Friedmann, he came back from Perm to Petrograd in the spring or summer of 1920. V. A. Steklov's diaries enabled us to fix the date. In his diary for 1919–20, there is an entry dated May 20, 1920: "At 11.30 p.m. Tamarkin and Friedmann turned up unexpectedly. They came from Perm in a special stove-heated freight car,[1] the ride took them twelve days. In Perm they were detained and searched, were about to have their provisions taken [let us recall that the Civil War was not yet over, and profiteers were severely dealt with – V. F.], but everything was sorted out. Friedmann wants to settle here for good and, of course, at the University ... He is very talented! Of course, he is asking for assistance. I said I would talk to [A. A.] Ivanov, but I will not make any arrangements. Let them act on their own. I may see Ivanov on Sunday."

Steklov's biographers note that behind the outward sternness there was a sympathetic soul. Thus, as soon as May 23, he writes: "At 1 p.m. there was an emergency session of the Pulkovo Committee. I told Ivanov that tomorrow Friedmann would submit an application to get a transfer here [i.e. to the University – V. F.] as a lecturer. He promised to arrange everything."

How did Petrograd in the latter half of May 1920 strike Friedmann? Let us recall that the Civil War was going on. The newspapers were publishing dispatches from the battle fronts. On the northern front, close to Petrograd, there was shooting in the Sestra River area. There were no changes on the Caucasian front and in the Maritime Territory in the Far East. The hot spot, or rather line, was the western front. There was fierce

[1] The freight car was loaded with books which Perm University was returning to Petrograd University.

fighting with the White Poles. On May 6 the Red Army abandoned Kiev, on May 14 its counter-offensive began, there was heavy fighting in the Chernigov area, a crossing was forced over the Berezina River, the town of Kanev was liberated, and a major battle was beginning at Vapnyarka. But the White Poles launched a counter-offensive in the Gomel area. On May 20, a rally was held in Petrograd: "The current situation and the White Poles in Soviet Russia." However, the city was gradually returning to peaceful life, as is shown by reports in the *Krasnaya Gazeta* (Red Gazette). Work was started in the docks; one can read appeals to save fuel next to reports about the regular deliveries of firewood to the city. On May 20, Chaliapin came back from a tour of Estonia, and rehearsals of *Khovanshchina* were resumed with his participation. Theaters and movie houses were open, a rowing race was held on the Neva. At 12 o'clock on May 20, on a gun signal from the Peter and Paul Fortress, a parade was held on Uritsky Square to celebrate "the holiday of Militia and Red commanders." One of the militia's routine tasks was to combat profiteering. There are reports about the nationalization of the book stocks accumulated in warehouses all over Russia, "the aim of which is to put an end to profiteering in books, which has reached a considerable scale of late."

Let us return to Steklov's diary. In the summer of 1920, one can see in its pages the names of Friedmann and Steklov's other pupils. June 29, 1920: "Tamarkin, Friedmann and Gavrilov turned up . . . We sat together until almost 1. a.m. Drank a whole samovar of tea. They brought more than a pound of salt! It's good!"

As early as July 12, the University Council elected A. A. Friedmann as a lecturer in mathematics and mechanics. He was to give a special course in applied mechanics (two hours a week) and to conduct tutorials with undergraduate students on definite integrals.

On August 6, 1920, having already settled in Petrograd, Friedmann wrote a letter to Ehrenfest. Paul Ehrenfest's first post-war letter to Petrograd came to the Physical Institute of the University – probably to D. S. Rozhdestvensky or K. K. Baumgardt. ("You can't even imagine what a joy it is for the Physical Institute to get your letter," writes Friedmann.) In his letter Friedmann outlines the results of his work in Perm. There are also a few lines about the new work he started soon after his arrival in Petrograd:

I have been working on the axiomatics of relativity principle, proceeding from two underlying propositions: (1) uniform motion remains uniform for the uniformly moving world and (2) the velocity of light is constant (identical in the moving and stationary world), and have obtained formulae for a world with one space dimension which are more general than Lorentz's transformations, and have an extra parameter. In a world with two or more space dimensions, the formulae of velocity composition (properties of the group of transformations) enable Lorentz's formulae to be obtained. This distinction between the one-dimensional and other worlds seemed droll to me. I have recently begun to reflect on what kind of arithmetic people would invent if the velocities available to them were close to the velocity of light. I would like very much to study the strong relativity principle, but have no time."

"I would like very much to study the strong relativity principle ..." This phrase is a prelude to Friedmann's cosmological investigations. There is no doubt that his interest in the "strong relativity principle" – as the general theory of relativity was called at the time – was the result of his contacts with theorists living in Petrograd, and especially with V. K. Frederiks and Yu. A. Krutkov – Alexander Friedmann could meet with them at the sessions of the Atomic Commission or at the University.

So, the last period of Friedmann's intense life and hard work began: Petrograd 1920–25.

At Petrograd University and colleges

On returning to his native city, Friedmann immediately plunged into work. By the autumn of 1920 he was already working:

(1) at Petrograd University – giving lectures and conducting tutorials;

P. S. Ehrenfest.

(2) at the Institute of Railway Engineering – giving lectures in aero-mechanics at the Faculty of Air Communications;
(3) at the Petrograd Polytechnical Institute, simultaneously in the departments of mathematics and mechanics of the Faculty of Mechanics;
(4) at the Naval Academy – in the department of mechanics;
(5) at the Mathematical Bureau which he organized at the Main Physical Observatory.

One should remember the acute shortage of specialists in Petrograd at the time in order to understand the reasons for this "superpluralism." And one needs to know A. A. Friedmann to understand that in each position he occupied he dedicated himself completely to the work exhausting himself and ruining his health, already shaken by wartime ordeals.

It seems only natural that Alexander Friedmann should have come to work at each of the above-mentioned institutions.

At the University he was no stranger, having been connected with it since 1906! His teachers were still lecturing there; his comrades of his student and postgraduate years, had started to work there. Initially Friedmann's teaching load was not great: two hours of lectures a week in applied mechanics and two hours of tutorials in definite integrals. This was not, of course, all that he was concerned with. Teaching reform at the University was long overdue. On July 2, 1920, there was a spontaneous rally at the University on this issue, which was attended by Maxim Gorky. Steklov wrote in his diary that many professors were sabotaging the reform as a whole; to counterbalance them, a special commission was formed to promote the reform. The situation was far more favorable with regard to reforming the teaching of exact sciences. On July 4, in Steklov's usual words, "Tamarkin and Friedmann turned up;" they discussed organizing a mathematical institute at the University and asked Vladimir Steklov to take this initiative into his hands. Steklov writes with bitterness that his relations with the university staff had become strained (besides, let us add, he was extremely busy at the Academy of Sciences at the time). "Nevertheless, if my students sincerely wish to do something – let them take my advice," he wrote. "I briefly outlined my point of view, saying that it was silly to think about a separate organization without raising [the question of] the general reform of the faculty, the more so now that it was imminent and being prepared by the Bolsheviks themselves ... I told them to write a suitable memorandum and to bring it to me at the Academy on Thursday ...

They left at 12.30," he concludes, and, as always, notes: "It's 2.30 a.m. The sky is clear, 19 °C, pressure 764.5 mm."

Tamarkin and Friedmann were making quick progress. On July 8, they brought to Steklov a draft project for a mathematical institute (which he undoubtedly took into account when he took charge of the Physico-Mathematical Institute of the Academy of Sciences). On July 10, the draft was discussed at the session of the mathematical section of the faculty, and on July 12, at the session of whole faculty. Unfortunately, the first attack launched by Steklov and his disciples was not a success; they failed to get support from the faculty and university administration. This is what Steklov noted in his diary on July 27: "They have retained this meaningless faculty in its present-day chaotic, shapeless and unprincipled form." But the steps they took did, albeit later (when Yu. A. Krutkov, V. I. Smirnov, V. A. Fock and many others got involved in this enterprise), eventually achieve their purpose.

Alongside the organizational work, the "particular" program of lectures was being prepared with the participation of Gunter, Tamarkin and Friedmann for institutions supervised by the Faculty of Physics and Mathematics, e.g. for the Petergof Natural Science Station and others.

And each time they met with difficulties at the University, Steklov's pupils would turn to him for support. Thus, in Steklov's archive there is a letter (of November 1924) addressed to him by V. I. Smirnov, N. M. Gunter, N. N. Gernet, B. N. Delone, Ya. D. Tamarkin and, of course, A. A. Friedmann, asking him to help retain at the University some of their students who had not been supported, for unknown reasons, by the members of the selection committee of the Faculty of Physics and Mathematics.

At the University, Friedmann was giving lectures, examining students. Prof. O. N. Trapeznikova, who attended these lectures, remembers that in those years the lecturers were assigned the task of issuing the students with ration cards for 150–200 grams of bread. How much attention and care Alexander Friedmann used to give to this! Dr. A. B. Shekhter (in 1924 she began writing a thesis on general relativity under Friedmann's supervision, completing it in 1928, after his death, under V. K. Frederiks' supervision) recalled the following episode. A female student felt terribly nervous at the examination in mathematics, and gave some inarticulate answer to the question. "Well," said Friedmann, "give me your examination card." He gave her the "credit" and asked: "Have you now calmed yourself? Everything is all right, isn't it? Then take your seat and answer the questions." And the student gave excellent answers to all the questions.

It was at the university seminars, as we have mentioned before, that Friedmann presented his epoch-making investigations in relativistic cosmology.

Friedmann was no stranger at the Institute of Railway Engineering either, having worked there already before the war! In the autumn of 1921, Alexander Friedmann took the position of Professor of Applied Aerodynamics. On November 14, 1921, classes were inaugurated at the Faculty of Air Communications (it should be remembered that there were no aviation colleges at the time – and one can imagine how much Friedmann knew about aeronautics and aeronavigational instruments!) The lectures in aerodynamics and aeronautics were given by the most outstanding expert in the field, and impressed the audience greatly! N. M. Gunter recalled that there were more lecturers than students among those who attended Friedmann's lectures in approximate calculations. Besides giving lectures in the above-mentioned courses, Alexander Friedmann conducted tutorials in various areas of mathematics, and took part in writing a book on problems in higher mathematics which went through five editions. For the first edition (1912), it is true, Friedmann and Tamarkin, according to the foreword, rendered assistance to its authors – A. A. Adamov, N. M. Gunter, Ya. V. Uspensky – in reading the proofs. In the three subsequent editions (the third one was published in 1924) Friedmann and Tamarkin were editors along with Gunter and Uspensky. finally, in the fifth edition (1931), the latest known to us, the cover and the title page bore the names of ten authors, including several mathematicians who have already been mentioned in this book: A. A. Adamov, V. V. Bulygin, A. A. Friedmann, N. M. Gunter, V. I. Smirnov, Ya. D. Tamarkin, Ya. V. Uspensky. One may suppose that the three-volume collection of problems in higher mathematics, by N. M. Gunter and R. O. Kuzmin, which was popular in the 1930s–50s and went through 15 editions, was largely based on the above-mentioned collections.

The course of lectures which was initially given at the Institute back in 1913 is likely to have served as the basis for a later book *Approximate Calculations*, which was written by Friedmann jointly with Ya. S. Bezikovich (a mathematician and A. S. Bezikovich's brother, who toward the end of his life, after the war, was Professor of the Urals Polytechnical Institute in Sverdlovsk). Its first edition came out in 1925, during Alexander Friedmann's lifetime, and the second in 1930.

Friedmann must have been invited to the Petrograd Polytechnical Institute by A. F. Ioffe. It was in the autumn of 1919 that Ioffe organized at that institute a new faculty: of physics and mechanics, which was to train physical engineers who were equally needed at research institutes and industrial plants. The faculty was to provide courses in physics, mathematics and mechanics. Ioffe was trying to involve in the work of the faculty the best of the Petrograd scientists. Among them were scientists already known to the reader (A. A. Adamov, V. R. Bursian, I. I. Ivanov, Yu. A. Krutkov, A. N. Krylov, I. V. Meshchersky), as well as others whose names have gone down in the history of Soviet science (P. L. Kapitsa, N. S. Kurnakov, M. V. Kirpichev, Ye. L. Nikolai, N. N. Pavlovsky, V. V. and D. V. Skobeltsyn, and A. A. Chernyshov). By the time of Friedmann's return to Petrograd there were already regular lectures at the faculty. Very few students of the new department were yet attending the lecture courses in quantum theory (Yu. A. Krutkov), the relativity principle (V. K. Frederiks), atomic structure (V. R. Bursian), and the theory of magnetic phenomena (P. L. Kapitsa). It is to be noted that among the organizers of the faculty was A. N. Krylov who knew Friedmann well by his work at the Institute of Railway Engineering, by his geophysical research, by the sessions of the Russian Physico-Chemical Society (of which Friedmann had been a member since 1911) and by his service in the army in 1914–16. It is not surprising, therefore, that it was through Alexei Krylov's recommendation that Friedmann came to the Faculty of Physics and Mechanics. A copy of this recommendation is kept in Krylov's archive in the Leningrad section of the Record Office of the USSR Academy of Sciences. Krylov begins his recommendation in the old fashion: "By request of A. A. Friedmann, I have the honor to present to the Faculty of Physics and Mathematics of the Polytechnical Institute a review of his research and teaching activity." This is followed by a concise list of what Friedmann had done before the war. The war-time events were known to Krylov from his own experience! We shall quote the relevant passages here, although they would perhaps be more appropriate in the chapter on the war years:

After the war broke out in 1914, Al. Al. [Alexander Alexandrovich] volunteered for active service as a bomber in the Air Force and took part in military actions as a pilot in Eastern Prussia and Galicia, and was decorated with combat awards. His abilities and knowledge drew the attention of the Army commander, and he was appointed an army physicist and transferred to the Central Directorate of the Air Force in Kiev, where he was given a high-ranking position, becoming responsible for organizing the entire air defense, being at the same time a leading figure in the meteorological service of the Army in the Field, the importance of which was immediately recognized with regard to gas attacks . . . During my eight-month administration [of the Main Physical Observatory] I had the opportunity to appreciate the organizational skill, energy and practical ability of Al. Al. Friedmann.

And now this is what Krylov wrote about Perm University:

Here he was able to organize a faculty of mathematics and earned such high esteem among his colleagues that he was elected Rector of the University [actually, a Deputy Rector – V. F.]. Thus, Alexander Friedmann combined in himself the qualities not only of an outstandingly talented scientist, but also of a practical organizer. His four-year work in the Air Force placed him close to technological tasks. His involvement in the work of the newly created Faculty of Physics and Mathematics will be particularly valuable, since he will be able to combine the comprehensive scientific side of the course in mechanics with its applied character, using a number of instructive examples from his own practice.

Let us note an aptitude for organizational work, and Friedmann's sober pragmatism – which irritated Steklov, but was approved by Krylov.

On August 2, 1920, the 10th session of the Council of the Faculty of Physics and Mechanics was held to elect new lecturers, A. F. Gavrilov standing together with Friedmann. Alexander Friedmann was elected Professor of Theoretical Mechanics by nine votes to one, and Alexander Gavrilov (unanimously) Professor of Mathematics.

We know about Friedmann's teaching at the Polytechnical Institute from the article by L. G. Loitsyansky and A. I. Lurie. The authors, famous Soviet scientists, considering themselves to be Friedmann's pupils, wrote: "During the five years of his work A. A. Friedmann created at the Polytechnical Institute a school of mechanics, and trained numerous pupils, who later developed the ideas and pedagogical principles of their teacher. The present-day [1949 – V. F.] areas of hydroaerodynamics and the dynamics and resistance of machines are rooted in Friedmann's school."

Among Friedmann's first pupils at the Faculty of Physics and Mechanics was G. A. Grinberg. Friedmann suggested that he should develop independently the topic which combined the two major themes of his own research of those years: hydrodynamics and relativity. Grinberg was to develop further the theory of elasticity and hydrodynamics in accordance with (and taking into account) relativistic principles. In 1925, he published three papers on this subject – these were significant scientific results of the first graduate of the Faculty of Physics and Mechanics, who is now a Corresponding Member of the USSR Academy of Sciences and the oldest (since 1919!) researcher of the Physico-Technical Institute of the USSR Academy of Sciences.

Loitsyansky and Lurie, who attended Friedmann's lectures at the Faculty of Physics and Mechanics, recall that he was fond of illustrating his lectures with problem solving – problems and their analysis often took up to half the time given to the course. Thus, in presenting the subject of external ballistics Friedmann not only developed methods of solving the differential equations in ballistics, but also taught the students the art of making up charts. He justly believed that the skills acquired by the students in solving problems would teach them how to think independently and therefore, having failed to finish the course within the allotted time, he would indicate the literature the students were to read by themselves. And the students successfully coped with the task the lecturer had set.

All courses in mechanics given by Friedmann were saturated with vector analysis, which was a teaching novelty in those years. In his course Friedmann also boldly introduced problems in the non-linear theory of oscillations, an emerging field of the classical area of mechanics. Using the lecture material on theoretical mechanics, his pupil L. G. Loitsyansky (at the time an assistant lecturer in the department of mechanics), prepared a book which was printed at the Naval Academy in 1924. One cannot but recall how Alexander Friedmann, as an undergraduate and graduate student, used to prepare similar books based on lectures given by his university professors! But in this case, on reading the text prepared by Loitsyansky, Professor Friedmann highly appreciated the tremendous work of his student and put Loitsyansky's name on the first page next to his own.

Ye. D. Devyatkova, who was an undergraduate at the faculty of Physics and Mathematics at the Polytechnical Institute in the early 1920s, found it necessary, in her article about A. F. Ioffe, to tell about the way mechanics had been taught at the faculty. She writes:

Our class was lucky enough to attend for two years (in the second and third years) A. A. Friedmann's lectures on point and system dynamics. He had just completed his theory of the expanding Universe, and presented it in a way easy for us to understand. Friedmann would always come to his lectures with a summary, or crib as he called it, into which, however, he never looked, covering all the blackboards in the lecture room, one by one, with equations. But one day he forgot to bring his notes. Having begun the lecture, he soon said that their absence made it impossible for him to deliver the lecture. He put on his coat and left.

Friedmann did not only read lectures. He was willing to give the students problems to solve and worked with the students who displayed interest in some of the areas he himself was concerned with. He also wanted to organize a seminar and suggested a number of topics for individual presentations. Unfortunately this initiative was broken off by his death.

The episode told by Ye. D. Devyatkova correlates with the one quoted above, told by A. B. Shekhter. Undoubtedly, there was some similarity in the characters of the young student and the young professor: a certain lack of confidence, displayed to a different degree. In the case of Friedmann, Pestalozzi's words are worth recalling: "Lack of confidence in oneself and one's abilities is a companion of talent."

Speaking about the Faculty of Physics and Mechanics, I would like to note that all three authors of the present book – also graduates of the faculty of Physics and Mechanics of the Leningrad Polytechnical Institute – attended lectures by Friedmann' pupils, and together with their fellow-students are proud that they have graduated from the institute where Alexander Friedmann used to teach.

Very little is known about Friedmann's work at the Naval Academy. A. N. Krylov taught there for many years (and studied there in 1888–90). Furthermore, it was in the summer of 1919 that he was elected Superintendent of the Academy. There is no doubt that Friedmann was invited to the Academy to give lectures, initially as a junior research assistant in the department of mechanics. We have already mentioned that Friedmann, in his turn, involved his university colleagues in the work of the Academy. One of them, A. F. Gavrilov, writes about Friedmann's lectures in mechanics at the Academy as being "most interesting." Gavrilov also notes the exceptional diligence with which Alexander Friedmann prepared all his lectures. He would always "write the texts in advance, thoroughly thinking out all the drawings, symbols, computational details, and order of magnitudes." The latter were needed by him for the approximate, "on the fingers," solution of various practical problems, and to estimate effects.

However, most of his time Friedmann seemed to give to the Main Physical Observatory. His work in the meteorology and physics of the atmosphere has been described in the previous chapters, and the last period of his work, as its Director, will be dealt with in the next.

Visit to Germany and Norway, 1923

On the initiative of the Main Physical Observatory, Friedmann was sent on a trip to Germany and Norway. In his 1925 autobiography he made a slip of the pen: he wrote that the trip was in 1922. This slip later got into many articles about him. In actual fact, he left for Germany in July 1923. Besides purely scientific tasks, Friedmann had an assignment to purchase books and instruments for the Observatory.

Very little was known about this trip, but the situation changed several years ago when the Leningrad section of the Record Office of the USSR Academy of Sciences received letters from Alexander Friedmann to

Natalia Malinina, many of which were written during his trip abroad in 1923.

These letters reveal a difficult period in Friedmann's private life. In 1921 he met Natalia Yevgenievna Malinina, a young worker at the Main Physical Observatory, who was working at the time on problems of Earth magnetism and, particularly, the Kursk magnetic anomaly.[2] She was a talented cheerful woman, keen on literature and art, personally acquainted with V. V. Mayakovsky, K. I. Chukovsky and other out-standing Soviet cultural figures. Friedmann's love for her was clouded by thoughts of his first wife Ekaterina, a person of high moral qualities, who lived entirely for the interests of her husband. Even after deciding to divorce his wife Ekaterina, Friedmann had pangs of conscience; some of his letters to Malinina are painful to read. In one of them (March 1924) Friedmann wrote that the only acceptable way out for him would be to part with life, but adds: "I can't commit suicide now – I don't have the courage." One may think the reason why Alexander Friedmann was suffering so much, was that both women were beyond reproach in their behavior in this delicate situation. In another letter to Natalia, after he had married her, he wrote about the inner conflict that was torturing him; he compares himself with a pendulum: "On my path, as a symbol of the extreme points of my oscillations, stood you and Ekaterina." It is justly said: "Don't read other peoples' letters!" but what if they are a source of important information?! We shall quote here the most interesting excerpts from Friedmann's correspondence with Malinina, trying to be as tactful as possible with regard to their memory.

Moscow, June 22, 1923. In Moscow there is the usual hustle and bustle: some call it life. The streets are full of vendors, traders, countless shops, pubs, beer-halls, restaurants and other eating-places, a procession of delivery men, cab drivers, overcrowded street-cars. For some reason, Daddy was not waiting for me in the first-class buffet as had been agreed, and I had to go to the railroad station twice to find out finally that he had not come at all. I'm worried about him.

"Daddy" was Vladimir Andreyevich Steklov, who had left Petrograd for Moscow by another train on business of the Main Physical Observatory. This was what Steklov's pupils called him between themselves and, sometimes, more gently – "dear daddy."

[After July 25, perhaps July 26, Moscow.] On Monday evening I was flying in an airplane (De Havilland D.H.9, speed 180km/h). It was exciting as always.

[2] N. Ye. Malinina (1893–1981), for many years Director of the Leningrad section of the Institute of Earth Magnetism, Ionosphere and Radio Wave Propagation of the USSR Academy of Sciences.

Moscow looked very beautiful from above (the altitude was 1,200 m). I could see that my heart had become much stronger, since it did not sink even slightly.

Glavbukhflot [the Main Accounting Office of the Air Force] has allocated 10,000 rubles in gold to the calculation of air ballistic charts. I'm very glad of that, because now I will be able to provide for the workers of my department for almost a year which, given the possible perturbations of the MPO, is quite important.

Friedmann involved A. F. Gavrilov in the calculation of air ballistic charts. In his autobiography Alexander Gavrilov notes that Friedmann and he, together with some other associates, were heavily involved in this important task in 1923–24.

The next set of excerpts is taken from letters written already from abroad. They are addressed to Natalia Malinina in the town of Shchigry (Kursh Province). Shchigry was only 60 km away from the town of Tim in which Friedmann, engaged to Valentina Doinikova, was living in 1908!

July 29, 1923, Berlin. One is struck in Berlin by contrasts – the luxury of the rich (who make our nepmen look like sucking-pigs), the problems of life for the middle class, and the workers' misery . . . Queues to buy butter, potatoes and sugar are growing from day to day. Berlin is undoubtedly on the eve of major events, as is the whole of Germany; it's about time they got rid of the bourgeois (philistine) rags and their "wise" materialist prejudices. At present in Berlin there is a wild currency orgy; recently (three days ago) a dollar was worth one million marks now it is worth four million and is expected to rise further, thus, within three days, the mark's value has dropped 500 percent; it does not affect those who have dollars, but for those Germans who live on their salaries it amounts to disaster and malnutrition, but malnutrition different from ours, malnutrition with shops filled to the top and with people eating to their hearts' content and enjoying life.

August 6, 1923, Berlin. Yesterday I was at the Staaken airfield and spoke with Kurt Wagener – he is a very interesting person, and he told me many curious things, yet to have a clear understanding of the state of German aeronautics one should live in Germany for at least a month; advances in aeronautics are little reflected in literature, if at all, it's difficult to get things published in Germany ... My further itinerary is this: I'm in Berlin till 8/VIII, on 8/VIII I go to Hamburg, where I stay till 11/VIII, on 11/VIII I return to Berlin; when in Berlin, I go to Staakenberg to the airfield, and once again to Potsdam; on 13/VIII I go to Lindenburg where I am to be until 18/VIII, on 18/VIII I return to Berlin and on 19/VIII I go to Christiania [Oslo – V. F.] and Bergen. About a week later, on 26/VIII I return to Berlin, on 27/VIII I go to Göttingen where I stay for three days until 30/VIII or come back to Berlin, or go to Thuringia, to Fulda, which has an airfield for gliders. Gliders are likely to play a big role in the near future in the use of the atmosphere, and it would be useful to find out about their organization.

In the next letter, without a date, apparently of August 8, in the train on his way to Hamburg Friedmann writes to Malinina that he met Ficker, Director of the Prussian Meteorological Institute (we mentioned his name in Chapter 5), and discussed with him the possibility of his working in Russia. Ficker wanted that very much. After Friedmann's death, V. A. Steklov, who had met Ficker during the Academy's jubilee celebrations in Leningrad, offered him the post of Director of the Main Geophysical Observatory. After long hesitation Ficker eventually declined this offer. To return to August 1923, they agreed with Friedmann on the publication of papers by Soviet meteorologists in a German journal. "If I accomplish something similar with respect to papers in mathematics and mechanics, 3/4 (of the task) of my trip will be fulfilled."

August 10, 1923, Hamburg. Today I talked with Alfred Wegener and learned many interesting and useful things. Wegener, as is customary with "great men," was reading nothing but his own work. Keppen is a charming old man, yet too old, and though he thinks clearly, he thinks slowly. Tomorrow I am to have a detailed conversation with Wegener.

In a letter written the next day we read:

Wegener told me a lot of interesting things about his work on the question of continent formation, ... his hypotheses are well supported by empirical data, the continents float in liquid magma, initially all the continents were together, then they gradually moved apart; such continental drift is likely to have a great effect on magnetic phenomena and probably on the course of history, yet this is only my opinion and not Wegener's. In Seewarte I was shown the operation of a wonderful machine which predicts flood tides, there is a great deal of ingenuity in its design.

Wegener told me that the meteorological department he is in charge of has no program of work; when I told him about the program of work of our department of theoretical meteorology he was very much surprised and was doubtful whether this program could be implemented at all. I answered that we were fulfilling it for the third year running.

On August 12, Friedmann returned from Hamburg to Berlin and on the same day wrote to Malinina:

I felt extremely depressed. It was particularly depressing to see the German forest: tree upon tree, standing in lines; everything is swept, with nothing scattered, all twigs are arranged in heaps according to size. Pah! This isn't a wood, it's a record office. What these infidels have turned God's forest into ... To tell you the truth, I like our life with all its lack of culture and all its shortcomings more than Germany's bourgeois life.

August 18, Berlin. I visited the Lindenburg Observatory, the reception was fascinating, I was invited to a full-dress dinner by Geheimrat Hergessel himself – felt terribly nervous, was constantly looking around so as not to make a blunder in table manners: I had not been invited to exquisite dinners for a long time. As a matter of fact, Hergessel and his wife, were awfully amiable and kind. They first gave me wine to drink, then cognac, then Chartreuse, and all this in great amounts. My previous experience in aviation enabled me to pass this drinking test with dignity. After dinner the Hergessels suggested we make an entry in their quest album (I was with Anna Bogdanovna Feringer). I wrote: "And to our teachers' health we raise this goblet," Anna Bogdanovna translated this into German. Ficker was there too. The same day (in the afternoon) before dinner I gave a talk in the Observatory on my work and the work of the department of theoretical meteorology. My German was, of course, horrible, but the talk was appreciated. Hergessel said a word of praise, noting that there is no department like the department of theoretical meteorology of the MPO anywhere. I also managed to get their agreement on publishing some of my colleagues' papers in the *Beiträge zur Phys. d. Fr. Atm.*

On August 25 Friedmann left for Norway where he was in close contact with Hesselberg, whom, by the way, he already knew through their joint work in Leipzig in the spring of 1914. On August 25, he wrote to Malinina:

The Hesselbergs gave me an exceptionally warm and cordial welcome, I have lunch and dinner at their place every day, besides, they take all possible care of me. Recently they presented me with warm gloves; Hesselberg's wife goes in for music, and I promised her to send some from Petrograd. We have discussed many things with Hesselberg, I am going to meet with Bjerknes in a few days, he is coming to Christiania, so I won't have to go to Bergen.

August 26: This morning I was with Hesselberg in the countryside, in the mountains, from which there is a fascinating view of Christiania. The Hesselbergs have two sons. The elder one, Ivar, whom I knew in Leipzig as a baby, already goes to school; the younger son, Peter, is an extremely merry and lively boy who never cries ... We have made great friends, and I draw him boats, steamships, airplanes, and houses. He is teaching me to speak Norwegian, it's fun to be with him – he thinks I must understand Norwegian, explains things to me and gets angry when I do not understand. I would very much like to have such a boy as my son! ... Hesselberg and I are planning to write a book, and a fairly complicated one, which will combine my methods (conditions of dynamic possibility of motion of a compressible fluid) and his methods (consideration of the order of magnitude of various elements). It's difficult to say if anything will come of it. We also have another plan: to write a book on dynamic meteorology, but this is, of course, a matter for the distant future.

Alexander Friedmann was not destined to have the distant future – the plans did not materialize. As to the near future, he implemented his intentions and got to Göttingen, from where he wrote on September 10: "Strolling along the streets of Göttingen I remembered Pushkin's words from *Eugene Onegin*: 'Vladimir Lensky with a Göttingen heart.' Silence reigns here, all the energy has clearly gone inward, into scientific work. Göttingen is buried in gardens, the houses are very small and some are very old architecturally; the street lighting is beggarly, standing in the main street I found Ursa Major and Ursa Minor without difficulty, and determined ENE."

In 1925, a popular-science book by the German engineer J. Ackeret, *The Rotor Ship. A New Method of Using the Power of the Wind*, was published under Friedmann's editorship. Ackeret was an associate of the well-known L. Prandtl, one of the founders of experimental aerodynamics, and Director of the Institute of Hydroaerodynamics in Göttingen. This university town interested Friedmann for at least two reasons: as a specialist in hydrodynamics he was interested in Prandtl, as a specialist in the theory of relativity and a mathematician, in D. Hilbert. Nothing is known about Friedmann's meeting with Hilbert, but he did have business contacts with Prandtl. Prandtl had supported the inventor of the rotor ship, A. Flettner, welcomed Ackeret's pamphlet and even wrote the foreword. It is highly probable that one of the outcomes of their discussion in Göttingen was Friedmann's decision to translate this interesting book into Russian. It may be asked what was so remarkable in this fact, a minor episode in Friedmann's biography. We have drawn the reader's attention to it, because in this case too (as with his work on the theory of the aircraft, its wing and lift) there is a direct coincidence of Friedmann's interest in the problem of a sailless ship with Einstein's similar interest. Einstein devoted a special article to Flettner's ship.[3] Is it not surprising that both men were equally attracted not only by the major areas of science and technology, but by more modest branches? True, Friedmann does not have any studies in the quantum theory of radiation, but during the period in question he carried out two investigations directly linked with the theory of atoms and quanta.

Alexander Friedmann was still in Perm when, on D. S. Rozhdestvensky's initiative, the Atomic Commission was set up in Petrograd under the State Optical Institute. Its first session was held on January 21, 1920. Among the members of the commission were the Petrograd physicists V. R. Bursian, A. F. Ioffe, Yu. A. Krutkov, A. I. Tudorovsky, V. K. Frederiks, V. M. Chulanovsky, and the mathematicians A. N. Krylov, N. I. Muskhelishvili, and a little later V. A. Fock. The commission concerned itself with atomic structure, developing the Bohr–Sommerfeld quantum theory for the case of complex atoms. This problem is a typical

---

[3] The article was translated into Russian and published in the journal *Izobretatel i Ratsionalizator* (Inventor and Rationalizer), No., 6, 1965, p. 17.

many-body problem, which is solved by various approximate methods (which have been developed, in particular, in related – but classical, not quantum – problems of celestial mechanics).

On May 1, at the session of the Commission, on D. S. Rozhdest-vensky's proposal, A. A. Friedmann was made a member of the Commission (soon followed by Ya. D. Tamarkin). Less than two months after his return to Petrograd, in July 1920, Friedmann presented to the Commission a solution to the problem of the motion of an electron revolving along the inner orbit (with respect to the outer – valence – electron) in a given field of the nucleus and valence electron. Friedmann had worked out this mathematically complicated problem, reducing it, as was characteristic of him, to a numerical one.

In 1924, the solution of the problem was published by Friedmann jointly with Tamarkin (who had presented his version back in the summer of 1920; the 1924 joint article of the two old friends and co-authors synthesized and developed their 1920 papers). Another question of quantum theory with which Friedmann and Tamarkin were concerned in the Atomic Commission was the problem of adiabatic invariants, which had been clearly formulated in the early 1910s by P. S. Ehrenfest and elaborated by T. A. Afanasieva-Ehrenfest, Ya. A. Krutkov, V. A. Fock *et al.* The paper by Friedmann and Tamarkin is kept in the records of the Atomic Commissions.[4]

In his 1925 autobiography Friedmann mentions what he did in 1924 in a single sentence: in the spring of that year he was sent to the Netherlands and took part in the First International Congress for Applied Mechanics held in Delft. At the Congress Friedmann first of all presented his and L. V. Keller's joint paper on mathematical and physical questions of a compressible fluid. The paper triggered off a lively discussion in which Prof. Hergessel also took part (the year before, as you remember, Alexander Friedmann had been his guest in Germany). At the congress, Friedmann reported about the investigations of his pupils N. Ye. Kochin and L. G. Loitsyansky, as well as of some of his colleagues. The congress was also attended by A. F. Ioffe, A. N. Krylov, N. M. Gunter, and Ya. D. Tamarkin.

On his way back home Friedmann again visited Berlin. On May 2, 1924, he wrote to V. A. Steklov from Berlin: "Everything went well at the congress, the attitude towards the Russians was wonderful; in particular, I was included among the members of the [standing] committee for

---

[4] The analysis of this paper and of Friedmann's 1920 work is the subject of a special article by T. I. Yefremidze and V. Ya. Frenkel prepared for publication.

convening the next congress. A. F. Ioffe's paper was especially success-ful.[5] N. M. Gunter's work aroused great interest in Lichtenstein who is now editing the journal *Zeitschrift für Mathematik*. Courant from Göt-ingen got interested in Tamarkin's work. Blumenthal, Karman and Levi-Civita got interested in my and my colleagues' work. If we have time I will gladly tell you about the congress in detail."

In the Netherlands Friedmann also undoubtedly met with P. S. Ehrenfest.

We are approaching the last year of Friedmann's life. The above-given information about his activities in 1920–24 by no means covers everything he did, even if one includes what has been said earlier about his work in the theory of relativity and meteorology. Putting aside the quality of his work, which reflected the growing maturity of Friedmann's talent, let us give a quantitative assessment. In 1920–24, out of the total of 51 titles listed in the biographical supplement to *Classics of Science*, about 20 articles were published (against 19 before 1920), and six books were written (two of them were published, though in 1925). We shall conclude this review by mentioning that Alexander Friedmann was not only editing the section on geophysics and meteorology in the first edition of the *Great Soviet Encyclopedia*, but also managed to write three articles for it (which were not included in the list of his works and were discovered by the present authors).

The first of them is "Arithmetic." Friedmann took an interest in the problems of arithmetic when a university student, and returned to them in some measure when he was writing his popular-science book on the theory of relativity. The article is perhaps most impressive in the historical introduction, showing the author's great erudition in the history of mathematics. There is no need to explain why Friedmann was invited to contribute to other articles – "Atmosphere"[6] and "Aerology." Alexander Friedmann did not live to see the publication of the Encyclopedia's Volume I, which came out in 1926. Its first page carried an announcement about the death of two contributors to the Encyclopaedia: M. V. Frunze (the editor of the military section) and A. A. Friedmann. The first of Friedmann's three above-mentioned articles appeared in Volume III, the other two in Volume IV.

Friedmann's life was filled to the brim with work – scientific, teaching and administrative. No wonder that so little is known about the way he

---

[5] Ioffe's paper presented at the Congress ("Plasticity and Strength of Crystals") was published in the *Proceedings* of the Congress, as, of course, was the paper of Friedmann and Keller.

[6] The article was written by Friedmann jointly with B. Izvekov and Ye. I. Tikhomirov.

spent his leisure time: in fact, he had none. The three weeks of his summer vacation that he was to have in 1925, the last year of his life, appeared to be the only rest he had throughout his "adult" years (in the town of Tim in 1908 he was not having a rest – he was working!). What did Friedmann read? We touched on this in Chapter 3, quoting V. V. Doinikova. We know, from Ekaterina Friedmann's reminiscences, about Alexander Friedmann's love for Dante ... And what about music? His father and grandfather were composers, his mother and aunt pianists. Is it possible that nothing was handed down to him? I have before me a letter dated February 15, 1971, written by Prof. N. N. Mirolyubov, one of the first graduates of the Faculty of Physics and Mechanics of the Polytechnical Institute. Nikolai Mirolyubov was not only a major expert in electrical engineering, but also a person of outstanding musical talent, and had a beautiful tenor voice. This is what he wrote: "That you decided to write about Friedmann is highly commendable. I knew him, since he was teaching us the basic course in theoretical mechanics. We had the deepest respect for him both as a scientist and as a person. I liked him also because he seemed to take a serious interest in music. I recall how surprised I was at first to see Alexander Alexandrovich and the mathematician Tamarkin sitting in the first rows at philarmonic concerts, with some books: later it turned out that they were following the score of symphonic pieces."

Happening to be with V. A. Steklov in Moscow in June 1923, Friedmann did not miss the opportunity to attend a ballet performance at the Bolshoi theater with his teacher. It is worthwhile to quote here Steklov's note in his diary dating to an earlier period, the one of December 26, 1910. That day he was visited by young mathematicians: "In the evening Tamarkin and Smirnov came, they played duets. I was singing under my breath. Smirnov left at 12.30, and Tamarkin at 3 a.m. Rather late!" Although the very frequent notes about the parties which Vladimir and Olga Steklov arranged for our mathematicians do not mention music specifically, there is no doubt that there were sounds of music in their apartment in Zverinsky Street.

11

The final year

Director of the Main Geophysical Observatory

In February 1925, A. A. Friedmann was appointed Acting Director of the Main Geophysical Observatory, and in June was confirmed in this post. It was another example of what N. M. Gunter was to mention in his speech in memory of Friedmann: "His practical activity was such that it could not fail to be noticed by everybody, no matter under whom he worked; we know that everywhere Alexander Alexandrovich very quickly rose to the highest administrative positions." Friedmann himself somewhat regretted the high appreciation of his organizational talent. In his letter to Steklov from Perm (of June 9, 1918) he complained: "My quick grasp of practical matters often renders me ill service, because my colleagues try to give me some responsible practical task; I flatly rejected administrative university posts, as to other kinds of work, I found it embarrassing to reject them at the very beginning . . ." As in Perm, in Leningrad Friedmann's resistance was broken and he plunged into public and administrative work. According to Gunter, Friedmann resurrected the Leningrad Physico-Mathematical Society: "The society began to work properly when A. A. became its secretary. He drafted the first charter of the society and got it adopted. Extremely busy, to be found only in his observatory study, he found time to attend the sessions of the board, he took upon himself the supervision of abstracts contributed to the *Zeitschrift*, advocated the publication of a journal during his visits to Moscow and was going to take part in editing it." In February 1925, Friedmann became the editor of the geophysics section in the *Great Soviet Encyclopedia*. And he never interrupted his research and teaching.

But the Main Geophysical Observatory took most of Friedmann's time and energy in 1925. Characterizing Friedmann's activities in his above-mentioned speech Gunter said that Alexander Friedmann was "the director who took this post with a prepared plan for a complete reform of the Observatory, and perhaps the only person capable of implementing this reform."

By the time Friedmann became the Observatory's Director, the Observatory and the Soviet meteorological service at large were in a difficult situation. Academician Krylov's resignation from the Director's post in early 1917 had weakened the traditional ties of the Observatory with the Academy of Sciences. After the October revolution, the Main Military Meteorological Administration, which had been headed by the Directors of the Observatory, was abolished. Inter-departmental strife which had been shattering the meteorological service in Russia did not disappear in the early postrevolutionary years. The report "On the state of the meteorological service in Russia and methods of its reorganization" written in late 1921 or early 1922 says: "Observers, getting from the Observatory no pay for their observations or a lower pay than from other bodies, send their observations only to meteorological departments of those bodies, denying them to the Observatory ... And an even more characteristic indication ... is a recent attempt of one department to set up a meteorological administration not of departmental, but of all-Russian importance, so that to save the Russian meteorological service from ultimate disintegration the People's Commissariat for Education had to quickly adopt a decree which confirmed the Observatory's historical right to supervise the meteorological service in Russia ..." This report, as well as a number of other documents concerning the work of the Observatory in the early 1920s, is kept in V. A. Steklov's archive which we have often mentioned already. Steklov, Vice-President of the Russian Academy of Sciences, concerned himself, despite the Observatory's formal "secession" from the Academy, with the Observatory's affairs till the end of his life. Steklov's archive has a copy of the decree of the Council of People's Commissars on the organization of a meteorological service in the Russian Federation, signed by V. I. Lenin on June 21, 1921. The decree assigned "the administration of the whole of the meteorological service in the RSFSR" to "the Main Physical Observatory (MPO) which is the central geophysical institution of the Republic," and "institutions and departments" were prohibited from "having a weather service separate from the MPO," and the existing "individual departmental organizations" were placed under the Observatory's control and were to be included "in the RSFSR's single Weather Bureau at the MPO." Having received full authority from the Soviet Government, the Observatory took active measures to re-establish the network of meteorological

stations. It became possible to save the major part of this network thanks
to the heroic efforts of the observers. Several years later, in May 1925, the
First All-Union Geophysical Congress noted in its special unanimously
adopted resolution:

The Congress especially notes the heroic efforts of the observers of the meteorological stations and observatories, who continued their work half-starving, sometimes under fire and very often in danger of their lives, who took measures to
protect the instruments and continue observations and who, along with the staff
of research institutes and observatories, contributed to the fact that the geophysical service, despite the hard conditions of the recent period, is now performing
serious and continuous observations and research, which are so important to
restore and meet the needs of the national economy of the Union Republics.

The successful restoration of the meteorological service was hampered
by the unfavorable situation at the Observatory itself. In the early 1920s,
it was headed by V. N. Obolensky, "a person without will who permits all
kinds of external interference in administrative matters and unchecked
spending of money by certain politicians and intriguers." This reference
was given in a report sent to Glavnauka – the Main Administration of
Scientific Institutions under the People's Commissariat for Education.
The report informed the Main Administration about a stormy two-day
meeting of the Observatory's staff, which took place on April 19 and 20,
1923. The meeting, chaired by a member of the Petrograd Soviet, Professor N. P. Kamenshchikov, demanded a "radical change" in the administration of the Observatory, organization of an auditing commission for
examining the financial activities of the Observatory, and elaboration and
approval of the Observatory's charter. On May 14, 1923, a commission
was set up to draw up the charter. It was headed by V. A. Steklov. Besides
him, the Academy of Sciences was represented by Full Members of the
Academy S. F. Oldenburg and A. Ye. Fersman, Narkompros (the
People's Commissariat for Education) by Ya. M. Hessen, and the
Observatory by N. P. Kamenshchikov, D. A. Smirnov and A. A. Friedmann. Friedmann was acting secretary of the commission. In the period
from May 23 to June 12, the commission had five sessions. In drawing up
the charter the commission was striving to legalize "the close connection
between the MPO and the RAS" (Russian Academy of Sciences). Paragraph 18 establishing the procedure for the election of the Observatory's
Director is characteristic in this respect. "The Director of the MPO shall
be elected from among persons known by their scientific work in the field
of physics and mathematics by the commission chaired by the President of
the Russian Academy of Sciences, which shall consist of three representa-

tives of section I of the RAS and three representatives of the Scientific Council of the MPO. The election procedure shall be established by the commission itself. The Director shall be elected for a three-year term and shall have the right to be re-elected. Should the Director give up his office, he shall remain Senior Physicist of the MPO. The election of the Director of the MPO shall be confirmed by Narkompros."

At its last session "the commission decided to ask V. A. Steklov, its chairman, to go to Moscow to explain in person and ensure the approval of the charter of the MPO. To assist V. A. Steklov, the commission decided to dispatch to Moscow the commission's secretary A. A. Friedmann." There are two notes dated April 19 and 21, 1923, which were written by Friedmann to Steklov and not included in their published correspondence:

Deeply esteemed Vladimir Andreyevich,

We are enclosing herein the draft of the explanatory note to the charter of the MPO. Would you be so kind as to introduce the appropriate changes in this copy which we managed to make on Monday.

Respectfully yours,

D. Smirnov and A. Friedmann.

Deeply esteemed Vladimir Andreyevich!

I am sending you several copies of the draft of the charter and of the explanatory note. I am taking with me a great number of copies; I am leaving for Moscow tonight.

Sincerely yours,

A. Friedmann

The charter was approved in 1924, and upon consultation with the Conference of Directors of Central Meteorological Institutions in Utrecht, the Observatory was renamed the Main Geophysical Observatory. Besides approval of the charter, a new head of the MGO was required: one who, in Steklov's words, had to command not only specialized scientific knowledge, but also to be an industrious, experienced organizer and administrator. Among the candidates considered were Steklov himself; Yu. M. Shokalsky, President of the Geographical Society, an oceanographer, who had been recommended back in 1917 by Krylov; P. P. Lazarev, a major Moscow physicist and geophysicist, who at the time headed research on the Kursk magnetic anomaly; and a young oceanographer, V. V. Shuleikin. It took a year and a half to solve the problem of finding a new Director for the MGO. Finally, on February 5, 1925, Order No. 556 was issued at the MGO:

In accordance with the decision of the Board of the Main Administration of Scientific Institutions (Glavnauka) of the People's Commissariat for Education of

January 27 of this year and the instruction of the head of Glavnauka, Comrade F. N. Petrov, of February 5 of this year, the Board of the Geophysical Observatory is relieved of its duties and, with effect from the same date, a new Board is appointed, consisting of Acting Director A. A. Friedmann, Deputy Director for Research Ye. I. Tikhomirov, Deputy Director for Administrative Affairs A. F. Wangenheim, and Members of the Board Ye. P. Dushevsky and S. A. Bastilov.

Acting Director Ye. Tikhomirov.

We can get acquainted with the program of the new Director of the MGO from his article "Prospects for the work of organizing the meteorological service" published in the first issue of the journal *Klimat i Pogoda* (Climate and Weather) started by Friedmann (the publication of this "popular journal of meterology and its applications" was warmly welcomed by the public, particularly by observers of meteorological stations, as shown by the column "Readers' Letters" which appeared in the second issue of the journal). Friedmann begins his article by formulating the most general tasks of meteorology:

The science of atmospheric phenomena (meteorology) is challenged by three tasks of exceptional importance for the country's economy. The first task is to make a diagnosis of the weather, i.e. to establish what the weather has been and is now in this or that region. The second task, based on the first, is essentially the task of foreseeing the weather, the task of prognosis – determining what kind of weather is expected in the near (short-term forecast) and distant (long-term forecast) future. Finally, the third task, undoubtedly the most important for practical work, though almost impossible, is the problem of artificial weather control, the art of "weather-making." One should note here that this third task is at present entirely impossible and fantastic, since to produce even the most insignificant atmospheric phenomena one needs energy of many hundreds of millions of horsepower working for many hours. Owing to the infeasibility of the third task it can be ignored and the whole attention of meteorology should be focussed on the first two tasks: diagnosis and forecasting of the weather.

Having analyzed the organization and technical condition of the meteorological observation service, Friedmann points out that to be able to "make a diagnossis of the weather" the Observatory should

(a) strive to achieve uniformity in the network of meteorological stations, as well as uniformity in methods of processing and publishing observation materials;

(b) give special attention to the organization of aerological stations and stations equipped with recorders;

(c) study and introduce new methods of observation, first of all observations of wind gusts and various non-instrumental observations.

Within the framework of an editorial by the head of a research establishment, Alexander Friedmann could not express his cherished dreams about probing into the atmosphere, and we learn about them from reminiscences of his colleagues. B. I. Izvekov wrote: "He was very much interested in the possibility of setting up a huge atmospheric laboratory, which he saw as a whole section of the Earth's atmosphere with a base area of several hundred or even thousand square kilometers, covered with a dense network of aerological stations. At fixed times a whole army of airplanes cross this section of the Earth's atmosphere in various directions and at different altitudes. The airplanes are equipped with recorders and thus ensure continuous registration of the dynamics of various meteorological elements at different altitudes ... Here Alexander Alexandrovich had a prophetic insight into the future, looking ahead of his contemporaries." It was, of course, unthinkable to elaborate on such a large-scale project in an austere, almost didactic, publication, so Friedmann expressed his idea in the most general form, stressing several times that observations should be performed "not only at the bottom of the air ocean, but throughout its entire depth."

Organizational, scientific and meteorological difficulties grow tenfold as one switches from diagnosing to forecasting the weather. And here Friedmann adopted the principle of a certain autonomy of regional centers and their links with the central weather bureau (the principle of "regionalism"). He adamantly opposed the division of the weather service between departments, believing that "departmental barriers" would be "totally disastrous for the meteorological service." As far as the research work is concerned, the Director of the MGO urges "an all-out effort directed toward studies of the laws governing atmospheric phenomena so that in the area of weather forecasting the relatively shaky basis should be replaced by a more solid scientific foundation."

Alexander Friedmann concludes his article by demanding that "personnel, equipment, and finances" be provided. This demand was to be heard from heads of scientific institutions many times in the subsequent six decades:

(1) It is necessary to create conditions which will enable the Observatory to have the required number of highly qualified researchers to perform its tasks.
(2) Those aspects of the Observatory's research ought to be enhanced which have clear and distinctly expressed aims serving to improve the national economy of the Soviet Republic.
(3) One should improve observational techniques and methods, as well as methods of processing and applying the results, to bring them closer to the present-day needs of geophysics and practice. In a number of cases this will require a discussion of relevant issues at international conferences, but

everything that does not require coordination with other countries should be implemented by the Observatory, inasmuch as this is dictated by the scientific interests and practical needs of our Union.

(4) The material base of the Observatory should be radically extended. It is necessary to raise the living standards of its staff and increase the funds allocated for its research and practical needs.

Only a vigorous implementation of these measures will make it possible to solve those major tasks of organizing the meteorological service which are so vital today and demanded by life itself.

Friedmann undertakes the implementation of his program with his usual energy. First of all, he takes "off the shelf" the project of sending an expedition to Yakutsk and Verkhoyansk to investigate the upper strata of the atmosphere, which had been drawn up back in 1912, but had not been implemented for lack of funds. The revised and extended project is submitted to the committee of the Academy of Sciences for the organization of the Yakutsk expedition. During the five years of its work it is expected to set up a geophysical observatory, and, in addition to the two meteorological stations operating in Yakutsk and Vilyuisk, to open others in Olekminsk, Buluk, Sredne-Kolymsk, Abyi, Suntar, Ust-Maisky, Nelkan, Yudomo-Krestovsky and the upper reaches of the Aldan River (in the area of goldfields). Besides four aerostations, the MGP was planning to provide the Yakutsk observatory with 12,555 rubles' worth of instruments, and books for a research and reference library. A famous polar explorer, at that time a senior physicist of the Main Geophysical Observatory, V. Yu. Vize, in his report submitted to the Presidium of the Yakutsk Committee, substantiated in detail the need for aerological observations near the pole of cold in Verkhoyansk, and the possibility of linking them with R. Amundsen's observations on the research ship *Maud*.

The task of installing equipment at the stations was assigned to the aerometeorological group headed by V. M. Chistyakov. The group left Leningrad on May 17, and on July 11, having got hydrogen for sonde balloons, started work on site. Within a few weeks equipment was installed at the stations in Olekminsk, Buluk and Ust-Timpton (on the Aldan River). Observation results were communicated daily to Leningrad by cable. The speed of the MGO's work was highly appreciated by the State Planning Commission (Gosplan). Its official (the Chairman of the Bureau of Congresses) A. A. Yarilov noted: "The Academy's Yakutsk expedition is a research institute for Yakutia, and the geophysical service has laid the foundation of this edifice."

Let us go back from the Yakutsk summer to the Leningrad spring of

1925. Two standing commissions are set up at the MGO. The actinometric commission headed by N. N. Kalitin was to raise to modern standards the studies of solar energy in the USSR. The Director of the Observatory took charge of the other commission – "the commission on the application of meteorology to practical purposes and on meeting the practical needs of departments." The commissions get down to work and Friedmann starts preparations for the First All-Union Geophysical (the Third Meteorological) Congress. He is elected vice-chairman of the convening committee of the Congress. Sixteen years abounding in wars and revolutions had passed since the Second Meteorological Congress. It was necessary to solve a number of scientific and practical problems of geophysics, and work out the plan of organizing the geophysical service under the new state system. The Congress had a complex organizational structure. Moscow University was to be the venue of the Organizational Congress, the Scientific Congress with six sections (general meteorology and aerology: actinometry; electrometeorology, atmospheric optics and acoustics; climatology, dynamic meteorology, and synoptics; earth magnetism, gravimetry and seismology; hydrology) and also the Conference on Drought. One of Friedmann's plans was to get the approval of the Congress for his program. As Friedmann's last letter to V. A. Steklov, Vice-President of the Academy of Sciences, shows, the "prime minister" of the major geophysical institution did not completely trust the meteorological "parliament," and tried to ensure in advance support for the plans of the MGO on the part of the Academy of Sciences and Gosplan. Friedmann's worries proved to be unjustified. He prepared thoroughly for the Congress; he convened beforehand a Conference of Directors of central geophysical institutions of national republics, and a session of the Commission of Directors of regional observatories of the RSFSR. The organizational principles elaborated by Friedmann became the basis of "the main principles of relations between the central geophysical institutions of the Union republics," adopted at the conference and reported to the Congress. The Congress approved the program of work of the Main Geophysical Observatory. The journal *Klimat i Pogoda* (Climate and Weather) wrote afterwards:

One of the resolutions of the Congress noted the correctness of the line adopted by the Main Geophysical Observatory with respect to the task of the MGO and the general meteorological service, groups of tasks, namely: (1) methodology of observations and their processing, (2) obtaining data for diagnosing the weather, (3) publication of diagnostic data, (4) investigating laws governing atmospheric phenomena, and (5) making weather forecasts, and in addition to these five groups its publishing work, the training of young researchers and the dissemination of meteorological knowledge among the broad masses of the working people – all this will enable the MGO to maintain its leading position among the meteorological institutions not only of the Soviet Union, but also of world meteorology.

At the scientific part of the Congress 235 papers were presented; most of them were of necessity poster presentations. The dissatisfaction of participants with the sections' work, which had been reduced because of the great number of organizational issues, resulted in a resolution on convening a scientific geophysical congress in Leningrad in the near future.

Ascent to the stratosphere

After the Congress the troubles and concerns of the Director of the MGO did not, in fact, become fewer; he stayed late at night in the building in the 23rd Linia of Vasilievsky Island. He came up with the idea of conducting an experiment on atmospheric vortices, a topic he had been concerned with for many years; not a laboratory experiment, but a full-scale one. "The investigation of atmospheric vortices near the earth's surface fails to achieve its aim, because the life of vortices is complicated here by a whole number of perturbing factors (uneven ground, buildings, the varied nature of the Earth's surface, etc.). Therefore, it is best to investigate vortices at a certain more or less considerable altitude where the earth's surface has no effect," – this is the way the editorial in *Klimat i Pogoda* (Climate and Weather) explains the need for a balloon flight. A spherical balloon, 1,437 m^3 in volume, was provided by the Military Aeronautics school. The flight was to be undertaken by the pilot Pavel Fyodorovich Fedoseyenko, the future commander of a fatal record-breaking flight in the stratosphere balloon *Osaviakhim-1*. The task of performing aerological observations was assigned by Director of the Main Geophysical Observatory Friedmann to the observer-pilot Friedmann. The order of Director of the MGO said: "Leaving on July 16 on official business in a balloon flight for scientific purposes, I temporarily entrust the duties of the Director of the Main Geophysical Observatory to the Assistant Director for Administration, Ye. P. Dushevsky."

One can see that high administrative rank could not change Alexander Friedmann's habit of doing unexpected things. One may wonder whether longing for the sky had awakened in him, or he had decided to enrich his "collection" of aeronautical methods: he had already flown aircraft heavier than air, airships too, only the balloon remained unmastered. Friedmann could have been excited by Professor N. A. Rynin's account of free flight. Nikolai Rynin was a lecturer at the Faculty of Air Communications of the Institute of Railway Engineering, where in 1920 Friedmann gave lectures in aeromechanics and was head of the preparatory courses. Rynin, who had set up an aerodynamics laboratory at the above-mentioned institute, was an enthusiast for flights into the atmosphere and beyond, and a friend of K. E. Tsiolkovsky, giving him various kinds of assistance. Rynin was Russia's record-holder for the altitude of balloon flights: in 1910, together with the pilot Odintsov he reached the 6,400 m mark. Attempts to surpass this achievement, undertaken after the end of the Civil War, failed: in 1921, an altitude of 5,225 m was attained, in 1922, 5,300 m. Fedoseyenko and Friedmann were destined to break the record of Odintsov and Rynin, and they succeeded. "At an altitude of 7,400 meters" – that was the title of the two aeronauts' accounts of their flight. Fedoseyenko's article appeared in the September issue of the *Air Force Bulletin*. Friedmann's story, written for the magazine *I Want to Know All*, was published in the double issue No. 2–3 of the journal *Klimat i Pogoda*. The accounts of the record-breaking flight were published in the magazine and the journal next to the obituaries. Friedmann's sudden death did not allow him to finish his scientific report on the expedition into the free atmosphere, but we know about the circumstances of the flight and the behavior of its participants in some detail from their accounts.

The preparation of the balloon for the flight took place on July 17. "The work is being performed by well-trained Red army men with loving care," wrote Fedoseyenko. "Everyone remembers that should he make a slight mistake, should he fail to spot a hole in a valve or mount bad springs, should he fail to spot a frayed string, the aeronaut is in mortal danger."

The car of the balloon was packed with the ballast, oxygen reserve, instruments and equipment. "The gas was not particularly fresh," the pilot noted, "yet we managed to lift up to 30 sacks of ballast." The volume of oxygen in a special "chest," three medical oxygen bags and five sondes was a thousand liters. With a consumption rate of 200 liters per hour per person, this amount of oxygen was enough for two or three hours of flying in rarefied air. Meteorological measurements were to be carried out with two barographs, an aneroid for measuring altitudes up to 5,400 meters, a mercury manometer, an Assmann psychrometer, and an instrument for measuring halo diameters. Like modern cosmonauts, the aeronauts of 1925 were to perform medical observations. Their scope was small, and the techniques extremely simple. They were to record their pulse rate and to register the accommodation of the eye muscles every thousand meters, using a special chart with a ruler. The last item of the program was "a microbiological experiment" – they were to take air samples with Petri dishes to determine bacterial content in the air.

Flight and landing safety measures were also quite simple. The aeronauts took with them two knives, binoculars, a clock, maps and two compasses. In case they flew at night, they took two electric torches with several batteries. In case they landed in an unpopulated area, they had matches. The possibility of coming down in water had also been foreseen: warm fur jackets were provided so that they could hold out three or four hours until help arrived.

The weather was not favorable for the flights. Cyclones were coming from the north-western parts of Leningrad, and in the north-east there was an area of high atmospheric pressure. "Thus," wrote Fedoseyenko, "Leningrad was in a system of saddle-shaped isobars, most unfavorable for the flight, since squalls and thunderstorms occur most often along the line connecting the saddles of isobars." On the morning of July 18, the sky was covered with rain clouds and stratus clouds. At 5 a.m. the air temperature was 13.2 °C, and by 7 a.m. it had dropped to 12.9 °C. Due to the cooling of the gas the balloon would not go up when released, and Fedoseyenko had to throw out two sacks of ballast one by one before the aeronauts, "having heard the last injunctions, greetings, wishes and instructions of Comrade Korrilov, the head of the School" began to ascend. The flight started at 7.10 a.m. The speed of the climb was 1.2 m/sec.

At an altitude of 230 m, the balloon entered the clouds, and at 450 m the clouds became so thick that the balloon could not be seen from the car. Fedoseyenko "measured" the cloud thickness by the length of the visible part of the guide-rope – a long rope hanging over the side of the car and designed to slow down the landing speed and to cushion the car landing (invented by Academician L. D. Zakharov in 1804). At an altitude of 450 m, the visibility was 1.5 m of the guide-rope's length. By 7.30, the balloon had reached the 1,650 m mark. The clouds began to clear up; first half the guide-rope, then the whole of it, became visible. At 7.45, at an altitude of 1,835 m, the aeronauts were delighted to see the sun through gaps in the clouds. But their joy was short-lived. Fedoseyenko wrote: "We were not destined to meet with the sun so soon. Man's struggle with Nature is not so easy . . . A huge white cloud on the left was slowly approaching us and enveloped us. It was the first time we have been in such a thick cloud: below the car the guide-rope could not be seen at all ..." Friedmann: "In the thick of clouds one can experience curious feelings and sensations. Total silence, total stillness, you see nothing and don't know what area you are flying over. Nobody sees you, and you can see nobody. Total isolation. At first, though, you hear from the ground the sounds of everyday life, steam-engine whistles, bells ringing, the crowing of roosters, the barking of dogs, etc. When you hear these sounds you feel more comfortable, but soon these sounds vanish too. Dead silence falls. This silence was, though, broken sometimes by strange sounds, as if somebody was hitting the balloon with a stick, like a drum – we first thought that the net was breaking, but then it turned out that the fabric was straightening out."

The aeronauts got adapted to the situation and started observations. At 2,000 m, they carried out the first series of medical observations. At 2,350 m, it started raining, the rain made the balloon heavier. Having got rid of part of the ballast, Fedoseyenko and Friedmann broke away from the rain clouds. At 2,650 m, they entered snow clouds. The rain gave way to "ice needles and snowflakes." They took the first microbiological air sample. The temperature dropped below zero. Fedoseyenko: "We can hardly keep the balloon from descending and are spending the ballast almost continuously in order to get higher." At an altitude of 3,370 m, the second layer of clouds ended, but nine minutes later they entered a third one, and it was only at 10.19, at an altitude of 5,200 m, that the balloon broke free of the clouds, and the blue sky and bright sun appeared above the aeronauts. Now the sunbeams would warm up the gas, lift the balloon and help to save ballast.

At about 6,000 m, oxygen starvation began to be felt. First they decided to have a "snack" with oxygen from the probes, but Friedmann tore two of them "through carelessness." They started breathing from the oxygen bags. Friedmann: "When we were restlessly inhaling the oxygen, an accident happened. The complete silence was broken by a deafening explosion. We looked up and saw that the whole of the balloon was shrouded in smoke. The thought flashed instantly to mind: fire." Fedoseyenko immediately went over in his mind all the options that could give at least some chance of survival. It turned out not to be a fire, but still a very unpleasant accident: the container with oxygen (the "chest") had burst, and the "smoke" was condensed moisture. Friedmann: "Knowing that our oxygen reserve was extremely limited, I would always refuse to inhale, saving the oxygen for the pilot." Fedoseyenko: "Professor

Friedmann is refusing to inhale oxygen, leaving it for me as the navigator, I have almost to force him to inhale oxygen, trying to convince him that we should support each other." Friedmann: "The selfless Comrade Fedoseyenko made me, through use of threats, 'feed' on oxygen, and I think that I owe my life, to a considerable extent, to these threats of his."

At 11.51, the aeronauts reached an altitude of 7,400 m. The aneroid had long been at "zero level," and the altitude was determined with the mercury manometer which shows 302 mm of mercury. At 12.30, to Friedmann's joy the balloon began to descend, but soon "jumped" again to 7,400 m. All in all, the balloon had stayed for some three hours above 6,000 m, and out of these three for over two hours above 7,000 m, while most of the oxygen stored for this altitude had gone. Friedmann: "All this time we were sometimes conscious, sometimes in a dead faint. These faints could be extremely dangerous, because we could have become too weak to perform the necessary manipulations to inhale oxygen. Had this happened, death would have been inevitable; two aeronauts, Crocé-Spinelli and Sivel, had thus died, after reaching, together with Gaston Tissandier, an altitude of over 8,000 m."

Despite oxygen starvation, Fedoseyenko performed a series of routine medical observations.

The aeronauts were helped by fleecy clouds which covered the sun. The balloon began to descend. Fedoseyenko: "13.00. The altitude is 7,100 m. Very difficult to breathe. The oxygen reserve is running out. 13.30. We are descending. Altitude 6,800 m, temperature − 18 °C. Down below in some places the earth is seen through gaps in the clouds. The pulse is getting weaker: 92–96. At an altitude of 6,500 m, we took a microbe sample." Friedmann: "The picture which unfolded before us when we were flying above the clouds was magnificent. Here and there above the flat cloud field there were high white hills gleaming in the sun. The hills formed a completely regular pattern − they were located at the corners of a net made up of diamond-shaped clouds. Between the cloud hills the air was entirely filled with ice crystals shining in the sun with all the colors of the rainbow; these little crystals seemed to radiate from one bright glistening center; at first it seemed to me that we were observing some electrical phenomenon, and it made me feel uneasy, the more so because on the horizon a huge thunderhead towered in the shape of a black anvil. And then I understood that the bright sphere, out of which ice crystals scattered in all directions like iridescent sparks, was an optical phenomenon, a kind of mock sun."

They continued to descend. The scorching sun made the aeronauts' faces burn with pain. At 16.17, an altitude of 5,200 m was registered. A new danger was approaching: they were about to enter the cloud stratum, and because of the cooling of the gas the fall could accelerate considerably, and by that time the aeronauts had only 56 kg of ballast (3½ sacks) left, not including "the economic ballast" – the provisions. But Friedmann was "absolutely confident in the skill of Comrade Fedoseyenko and was absolutely calm." And the navigator counted on the cloud hills which the professor admired: "Like faithful bodyguards, they will cushion the descent with their vortices, not letting us go down too fast." By 17.00, the aeronauts had got down to 3,800 m, having spent only half a sack of ballast.

Through thinned-out clouds they saw the Earth's surface: a coastline and a vast expanse of water. Was it Lake Ladoga or the Gulf of Finland? It soon turned out to be Lake Ilmen. The danger of "landing" on water became real, and it occurred to Friedmann that the safety jackets would not help. To slow down the descent and fly across the lake, Fedoseyenko used two sacks of ballast. They loosed the anchor cable, and Friedmann had to hold it back: "I had to put my fingers into the loop, because I could not hope to hold the cable in the hand, and at the same time it was not heavy enough to break my fingers. It was very painful, and my hand was hurting me for two weeks after that." But in the meantime, they were thinking about another thing; the lake was left behind, but Fedoseyenko and Friedmann did not know on what territory they were descending – Soviet Russia or bourgeois Estonia. "If they were fellow-countrymen, we would be given a warm and friendly welcome, there are friends of the 'Aviaradiokhim' everywhere. If they were not Soviets, it would mean imprisonment, even if temporary, and long conversations."

At an altitude of 1,000 m, they ran out of ballast. They descended to 300 m. They touched the ground with the guide-rope. They dropped the anchor and the balloon jumped up. But the anchor caught on a bush, and the balloon resumed its descent. Peasants ran up from all directions; the aeronauts, having become convinced that they were among their fellow-countrymen, began shouting through a megaphone that the "welcome committee" should catch the balloon. At 17.31, the aeronauts landed on Mshinskoye Field, 1 km south-east of the village of Okoroki.[1] Fedoseyenko and Friedmann found themselves in a remote corner of Novgorod Province, 100 km from the nearest railway station and 55 km from the nearest steamer wharf.

Soon after the completion of the flight Friedmann gave an interview to

[1] In the Mozheyevskaya *volost* of the Demianovsky *uyezd*.

the newspaper *Izvestia*. It has much in common with the above-quoted
more detailed account, published posthumously. There is, however, one
amusing detail omitted in the magazine version of the account. The
balloon landed in a field where peasants were working. This was so
sudden and unexpected that several women fainted! But youngsters
working in the field reacted to the "extraterrestrials" in a quite normal
way: they helped the aeronauts to put away the balloon and heard an
improvised lecture about their flight. On returning to Leningrad, Friedmann started correspondence with the Okoroko branch of Komsomol
(the Young Communist League). The return to Leningrad took place on
July 21, after "a rather tiresome and dull trip." The same day Friedmann
went for a short business trip to Moscow. He returned to Leningrad for
only a few days and on July 27 went via Moscow to Feodosia, where
Natalia Malinina was already waiting for him. A month before, on July 1,
1925, he had written her from Leningrad: "Now everybody is gone from
the Observatory, and I am alone among the statues and portraits of my
great and small predecessors, my soul after the day's bustle is becoming
calmer and calmer, and it gives me joy to think that thousands of miles
away the beloved heart is beating, the gentle soul is living, the new life is
growing, . . . the life whose future is a mystery, and which has no past."

Let us now turn once again to Friedmann's family life. In January 1923,

Alexander Friedmann and Natalia Malinina (left); after 1923.

he presented Natalia Malinina with his newly published book *The World as Space and Time*, with the inscription: "To highly esteemed Natalia Yevgenievna Malinina from the author."[2] Later that year, as has already been mentioned, she became his wife. During their short life together the happiest time was the period in question: July–August 1925. They were together again, Friedmann was merry and healthy, "he was jumping on the rocks like a boy," as one of his colleagues recalled. They hoped to live a long and a happy life. They even had a religious wedding ceremony in the Crimea, though both were far from religious. "Just to make it stronger," Friedmann said to his wife (she told this to her sister Sofia).

And many years later, in 1965, Natalia Malinina, restoring her inter-rupted acquaintance with Kornei Chukovksy, wrote to him (the draft of the letter has been preserved): "In 1923, I married Professor Alexander Alexandrovich Friedmann, a mathematician, a modest person, immersed entirely in science. He found an error in Einstein's calculations, and Einstein agreed with him later. Unfortunately, he died very early – in 1925. His son did not inherit his father's mathematical talent." Fried-mann did not live to see the birth of his son, to see the emergence of that new life, about which he wrote in such a touching and poetic way to his wife in July. Friedmann's son, the third Alexander Alexandrovich in their family, used to say to his mother when he was a teenager: "I want to be an ordinary person." He entered university during the war, served in the Army as a driver, and worked as a driver in Leningrad after the war, until his death in 1983. He had no children.

Friedmann returned from thè Crimea to Leningrad, and Malinina went to Shchigry, where she had work to do in connection with the geomag-netic station. On September 17, in a train on her way to Leningrad, she learned from a newspaper about her husband's death.

Last days

Friedmann returned to Leningrad, and on August 17, he was already at the Main Geophysical Observatory, and was there (judging by the Observatory's record of his orders) almost every day until August 23. And then, suddenly, he felt unwell. On September 2, he signed an order transferring his duties to Deputy Director Ye. I. Tikhomirov. The order was signed with an already shaky hand, apparently in bed. The doctors diagnosed typhoid fever. He had contracted it on the way home – an absurd accident. Alexander Friedmann was put into the Pervukhin

[2] This copy has been preserved by Solia Yevgenievna Malinina, Natalia's sister.

Hospital in Kamennoostrovsky Prospect. Before his illness, he had been busy with most urgent matters. V. A. Steklov had got him involved in the editing of A. M. Lyapunov's work *On Equilibrum Figures of an Inhomogeneous Rotating Fluid* which was to be published posthumously. Together with his pupils Friedmann was verifying each formula. They discussed the work with Steklov late at night – this time suited both of them, since both of them were hard-pressed for time and as a rule did research work at night.

The bicentenary of the Academy of Sciences was to be marked soon. Foreign scientists had been invited to take part in the celebration. The MGO was also to receive foreign guests. On his trip to the Netherlands, Germany and Norway, Friedmann had re-established the scientific links interrupted by the war and was going to strengthen them during the celebrations. The program for foreign scientists included visits to the Main Geophysical Observatory and its branches in Slutsk (now Pavlovsk): the Magnetic-Meteorological and Aerological Observatories. The Director was not among those who welcomed guests, and they were many – the meteorologists Meinardus and Ficker from Germany, and Melander from Finland; the famous physicists Planck, Raman, and Tanakadate; the mathematicians Levi-Civita, Kneser and Chapelon; the geologist Lee.

Meteorologists were interested in the Observatory more than anybody else. On September 11, the Director of the Prussian Meteorological Institute, Professor Ficker, published a very favorable assessment of Soviet meteorology in the *Transactions of the AUEC* (All-Union Executive Committee):

As regards the USSR, since the war the network of meteorological stations has been considerably expanded. What is interesting is that the network of Russian observation stations set up by the great Wild has not only been fully preserved, but has been considerably expanded to high northern latitudes and in Mongolia, for which world meteorology should be particularly grateful to the Union of Republics.

The Leningrad Main Geophysical Observatory, by increasing its staff, has managed during this time to become the leader in two directions in scientific meteorology. The method of mathematical calculations which is well developed here made it possible to set up a special department for theoretical meteorology, which at the present time is not to be found in any meteorological institute of the world.

At present, the Leningrad Main Geophysical Observatory is concerned with developing methods which will make it possible to forecast the weather for long periods of time. The problem is not, of course, fully solved, but I can say the Russian meteorology has come closest to solving it.

It thus follows from what I have said above that Russian meteorology is at present playing a leading role in world meteorology. In the interests of world development of this science I express a wish that the USSR should also organize aerological investigations using balloons and kites.

Friedmann was not destined to rejoice either at this recognition of the achievements of the MGO by the foreign specialist, or at the prizes awarded in October by Glavnauka to staff members of the Observatory – the senior physicist Ye. S. Rubinstein for the monograph *On Air Temperature in the European Part of the USSR*, the physicist M. I. Goltsman for new geophysical instruments and Friedmann himself for his *Hydromechanics of a Compressible Fluid*. Weakened by the deprivations of the war years and his exhausting work, he failed to throw off the disease, and on September 16, Alexander Alexandrovich Friedmann died.

The announcement of his death appeared in *Pravda*, *Izvestia*, and Leningrad newspapers. The evening issue of a Leningrad newspaper had an interview with Dr. Buchstab, who had been treating Friedmann, from which we learn that early in the morning on September 16, a stomach hemorrhage started. Alexander Friedmann was running a very high temperature and was delirious. His condition was described in the newspaper in a surprising way: "The delirium of the deceased was extremely characteristic: he was speaking about students and lectures, remembered the flight, was trying to make various calculations. Sometimes, he seemed to be giving a lecture."

The death of the young scientist, who had shown such great talent in his last years, shocked the scientific community. Friedmann's associates, his teachers and pupils, his foreign colleagues, were united in the feeling or irretrievable loss.

On September 17, the civil funeral was held in the Main Geophysical Observatory. Acting Director of the MGO Ye. I. Tikhomirov spoke about the impact the scientist had had on the observatory and geophysics at large. Rector of Leningrad University V. B. Tomashevsky paid tribute to Friedmann's memory. B. I. Izvekov, a researcher of the theoretical department, the learned secretary of the Observatory, reminded those present of the story of Alexander Friedmann's life.

On September 18, the funeral procession, headed by the President of the Academy of Sciences, A. P. Karpinsky, and the Academy's Permanent Secretary, S. F. Oldenburg, set out from the hospital for the Smolensky cemetery. The coffin was borne by students and colleagues from the Observatory and Leningrad educational establishments. The procession was accompanied by the students' band of the Polytechnic Institute.

Those who had recently cooperated with Alexander Friedmann in his diverse work came to pay tribute to him: a member of the board of the MGO, A. F. Wangenheim, who had now to write by himself a paper on the Neva floods, which he had planned to write together with Friedmann; Professor of the Polytechnical Institute I. V. Meshchersky, in whose department Friedmann used to give lectures; an instructor at the same institute, L. G. Loitsyansky, at that time Professor Friedmann's assistant, who was to continue his cause and to set up a department of hydro-aerodynamics; the pilot P. F. Fedoseyenko, his co-navigator; and Professor N. A. Rynin, the predecessor of Fedoseyenko and Friedmann. V. Osinsky spoke on behalf of the undergraduate students of the Institute of Railway Engineering, and D. I. Itskovich on behalf of the Communist party organization of the MGO.

Obituaries were published in Soviet and foreign journals and magazines: in *Klimat i Pogoda* (Climate and Weather), *Meteorologichesky Vestnik* (The Meteorological Bulletin), *Zhurnal Geofiziki i Meteorologii* (The Journal of Geophysics and Meteorology), *Uspekhi Fizicheskikh Nauk* (Advances in Physical Sciences), *Vestnik Vozdushnogo Flota* (The Bulletin of the Air Fleet); in the *Meteorologische Zeitschrift*; in *Nature*. "Only now that he is no more, that all search for a successor to this versatile scientist and outstanding administrator is in vain, only now is everybody becoming aware of the immensity of the loss," wrote A. F. Wangenheim in the second issue of *Klimat i Pogoda*, the very journal in whose first issue Friedmann had published his editorial several months before. Both in this journal and in the *Bulletin of the Air Fleet*, the obituaries were next to the accounts of the record-breaking flight accomplished by Friedmann himself and the pilot Fedoseyenko.

Leningrad scientific societies held sessions dedicated to Friedmann's memory. On October 23, staff researchers of the MGO, and members of the meteorological commission of the Geographical Society and geophysical section of the Russian Society of World Studies, met together at the Geographical Society. Ye. I. Tikhomirov read a paper on Friedmann's activities at the Observatory. Friedmann's associates L. V. Keller, B. I. Izvekov, P. Ya. Polubarinova and N. Ye. Kochin contributed a paper "A. A. Friedmann's investigations in dynamic meteorology" (the paper was presented by Izvekov). The MGO researcher M. A. Loris-Melikov spoke about Friedmann's work in the theory of relativity. The paper given by S. N. Troitsky was on Friedmann's work in aviation and aeronautics, and K. I. Vasiliev's paper was on Friedmann the teacher.

The Leningrad Physico-Mathematical Society dedicated two of its sessions to Friedmann. On September 26, the chairman of the society, N. M. Gunter, delivered a speech "In memory of A. A. Friedmann" (excerpts from it are quoted at the beginning of this chapter). Gunter attributed all Friedmann's theoretical achievements to mathematics:

In Alexander Alexandrovich, the outward shell of a person attending to mundane practical applications of an emerging science concealed the noble soul of an investigator of eternal questions of the Universe, and the noble face of a priest of pure knowledge.

The whole of his work is purely mathematical in essence. Many practically important things have been obtained by him because, having begun at the outset to learn how to reject, he did learn how to do it, and did it intelligently.

The connection between meteorology and the relativity principle becomes clear: in his independent studies of the relativity principle Alexander Alexandrovich was only doing essentially the same as he was doing in meteorology – investigating solutions of a system of partial differential equations.

On November 10, a joint session of the Physico-Mathematical and the Physico-Chemical Societies was held at the Physical Institute of Leningrad University. V. I. Smirnov spoke about Friedmann's mathematical investigations, I. V. Meshchersky about his work in hydrodynamics, and V. K. Frederiks about his work in cosmology. B. I. Izvekov again presented the paper on Friedmann's work in theoretical geophysics.

Letters and telegrams expressing condolences arrived at the Main Geophysical Observatory from all major geophysical institutes of Europe and America. Everding, President of the International Meteorological Committee, wrote: "Friedmann's death is a loss not only for the Main Geophysical Observatory, but for the whole of meteorology." "The Leipzig Geophysical Institute," wrote its Director Professor Weickmann, who had replaced V. Bjerknes after the latter's return to Norway, "has always been proud to call Professor Friedmann its staff researcher." "And for me personally," Weickmann added, "the deceased was a valuable companion and friend, whom I respected deeply for his high human qualities." T. Hesselberg, who had also worked with Friedmann under Bjerknes and was now Director of the Meteorological Institute in Christiania (Oslo), also shared his reminiscences about their personal friendship and joint work. Telegrams were also sent by English colleagues – "rivals" in studies of the vertical temperature gradient and the cyclone model – E. Gold and N. Shaw. The chairman of the International Commission on the Study of the Upper Strata of the Atmosphere, Sir Napier Shaw, wrote in his letter: "In the past year and a half, I came across Professor Friedmann's name so often that I naturally began to consider him one of the most industrious and successful investigators of atmospheric dynamics. I remember vividly the huge number of papers he brought to the International Congress for Applied Mechanics at Delft in April 1924, and evidence of his fruitful work has kept coming ever since. I learned with deep regret that death had terminated so early so promising a life."

Friedmann's pupils prepared for publication the manuscripts he had left. The fifth volume of *Geofizichesky Sbornik* (Geophysical Collection) contained the above-mentioned memoir on the theory of motion of a compressible fluid, the lecture "On approximate conditions of dynamic possibility of motion of a compressible fluid" edited by N. Ye. Kochin, as well as his joint work with P. Ya. Polubarinova on the motion of vortex sources. Alexander Friedmann's pupils continued work on their teacher's legacy, and it was completed in 1934, when *The Hydromechanics of a Compressible Fluid* was published with new results added, which had been obtained by B. I. Izvekov, I. A. Kibel and N. Ye. Kochin.

It can be noted here that the Academy of Sciences established the A. A. Friedmann Prize for outstanding contributions to dynamic meteorology. The winners have so far been I. A. Kibel (1972), Ye. N. Blinova (1979), M. I. Yudin (1981) and A. M. Obukhov (1984) (the reader has already come across these names in this book), and also G. I. Marchuk (1975), who is now President of the USSR Academy of Sciences.

12

Friedmann's world

Friedmann's theory discovered the most grandiose natural phenomenon, the cosmological expansion. Having passed through a rigorous checkup in the discussion with Einstein, it very soon found proof and confirmation in astronomical observations.

This concluding chapter of the book deals with the astronomical investigations that set up an observational basis for the science of the Universe; it discusses the further development of Friedmann's legacy and the ideas of modern cosmology, a living and rapidly developing science that eagerly absorbs the latest findings provided by astronomical observations and the brightest ideas produced by fundamental physics.

Galaxies and the Universe

Speaking of stars in his 1917 cosmological article (see Chapter 9), Einstein referred to them as the basic elements constituting the Universe. That was the period when the astronomical concept that our stellar system was unique enjoyed wide popularity. It was assumed that the Milky Way with all its stars, the Sun being one of them, was the whole Universe. Einstein too may have believed this. In the early 1920s, however, the astronomical picture of the world was altered drastically. The leading role in cosmology turned to be played by another class of astronomical objects – the nebulae.

Nebulae had long been known to astronomers, and they were regarded, within the concept of the unique stellar system, as comparatively small gas clouds floating among the stars. The true nature of the nebulae was established when the first major telescopes appeared. Three hundred years before, Galileo had invented a telescope and used it to see individual stars in the Milky Way. He proved therefore that the whitish strip

215

encompassing the whole sky consists of stars that the naked eye cannot distinguish. And now in just the same way, it became possible to see stars in nebulae. It turned out that these fuzzy specks were in reality gigantic stellar systems far away beyond the Milky Way. The discovery was made in 1923 by the American astronomer Edwin Hubble. Using the 100-inch telescope at the Mount Wilson observatory, he succeeded in obtaining exceptionally high-quality pictures of several spiral nebulae, in which individual stars could be clearly discerned. These spiral nebulae (later they came to be called galaxies) included the famous Andromeda Nebula, which is larger than, but very similar to, our own Galaxy. As Einstein wrote later (in his 1935 review of R. Tolman's book *Relativity, Thermodynamics and Cosmology*), galaxies "were destined to complement our knowledge of the structure of the space–time continuum." The Universe appeared in astronomers' eyes as a world of nebulae and galaxies rather than a world of stars.

Six years later, in 1929, Hubble made a new discovery. He established that galaxies were not at rest in space, but moved quite quickly. They were mostly moving away from us rather than towards us.

The motion of galaxies reveals itself in the famous "red shift:" the spectral lines of receding galaxies proved to be shifted perceptibly toward longer wavelengths, i.e. toward the red end of the spectrum. The pheno-

Edwin Hubble.

menon can be naturally interpreted as the Doppler effect, i.e. a change in wavelength when there is relative motion of the source and receiver. If the source and receiver approach each other, the received wavelength shortens, and if they move away from each other, the wavelength increases. The change of wavelength is greater if the relative speed is greater.

The remarkable discovery of the red shift had been preceded by other discoveries. In 1912–14, observations on nebulae were made by the American astronomer V. Slipher, who noticed their very fast motion. In particular, he demonstrated that the Andromeda Nebula is approaching us with a speed of about 300 km/s. Many other nebulae proved to be receding, again at a speed of several hundred kilometers per second. In 1917, these data attracted the attention of de Sitter, who saw the significance of the fact that most of the galaxies move away from us rather than towards us.

The English astrophysicist Arthur Eddington noticed it as well. In his book *The Mathematical Theory of Realtivity* (1923) he wrote: "One of the most perplexing problems of cosmology is the great speed of the spiral nebulae. Their radial velocities average about 600 km. per sec. and there is a great preponderance of the velocities of recession from the solar system." Eddington referred to Slipher's data on 30 moving nebulae.

Slipher's work was noticed in Petrograd. The journal *Mirovedeniye* (World Science) reported in 1923 in its "News in Astronomy" section: "According to the results obtained by the Lowell Observatory through processing Slipher's observations, spiral nebula No. 936 in Dreyer's catalog is moving away from us at a speed of 1,300 km/s. Spiral galaxy No. 584 of the same catalog has a speed of 1,800 km/s. Recall that the Earth's orbital velocity is only 29.76 km/s." The report was headed "Fast-moving nebulae."

In 1929, Hubble extended Slipher's list by data on another 18 galaxies moving at speeds of about 1,000 km/s. Hubble saw precisely the role and place of galaxies in the Universe, and in this new and not yet abundant material he could perceive a certain regularity in the galactic motions: except for some very close galaxies, all show the red shift and are therefore receding from us. It was proved that the speed of a galaxy moving away from us is greater, the farther is the galaxy from us. This dependence is called the Hubble law (or the red-shift law), and it has a very simple mathematical form: if v is the speed and R is the distance, then

$$v = HR.$$

H is called the Hubble constant; it is the same for any galaxy wherever it might be in the sky.

In 1931, Hubble received data on the motion of several galaxies, whose speeds reached tens of thousands of kilometers per second. The Hubble law proved to be very accurate for those galaxies (more accurate than for closer stellar systems). Generally, it reflects large-scale properties of the world, and the domain it applies to covers large volumes of the Universe.

Hubble's discovery became a direct astronomical proof that the Universe is non-static, or non-stationary. The expanding Universe must look exactly as Friedmann saw it. The Hubble law of velocities also follows from Friedmann's cosmological theory. The relative velocity of two bodies in the expanding world (meaning, of course, the sort of bodies that make up the Universe) is always proportional to the distance between them.

It was a mystery to some people: why should the galaxies recede from the Earth or from our Galaxy? However, this is only an illusion. As a matter of fact, we do not occupy a central position in the world. The Hubble law, implying a linear dependence of the relative velocity of two bodies on the distance between them, is valid wherever in the Universe the observer might be. (In order to prove this, it suffices to combine the Hubble velocities when passing from one observer to another.) Weyl once remarked that this fact reveals the homogeneity of the Universe, i.e. the equality of every "site" in it. As we can recall, such considerations of symmetry underlie Friedmann's theory: they have been used as the initial premises. Therefore, the homogeneity of the Universe can be proved by the Hubble law. But the law also indicates that the Universe is isotropic, that every direction in it is equal, because the law is valid for all galaxies regardless of their position with respect to us.

The homogeneity of the Universe follows from the uniform distribution of matter over its entire volume. Homogeneity, uniformity of distribution of galaxies, can be proved not only by the Hubble law, but also by direct calculations for these stellar systems. Hubble was the first to present reliable astronomical data. Adding up the number of galaxies in greater and greater volumes, he established that the number of galaxies increases in direct proportion to the volume. Therefore, the density of their distribution is everywhere the same, it is uniform. It goes without saying that here just as in the case of the law of velocities, only large-scale distribution of galaxies is taken into account. As to comparatively small volumes, where there are only a few galaxies, matter is distributed very non-homogeneously: it is mainly concentrated in the galaxies themselves and the surrounding space is almost empty. However, if larger volumes are taken, containing numerous galaxies in them, we can neglect these comparatively small non-uniformities in matter distribution and find out that in all those large volumes the average density of matter is the same.

The limit of observation at the Mount Wilson telescope was of the order of several hundred million light years. Hubble calculated that within a volume with a radius of 3×10^8 light years ($c. 3 \times 10^{26}$ cm) there were a hundred million nebulae. Apart from a certain tendency to cluster in space, they are distributed quite uniformly over the whole volume. According to Hubble's first estimates, the average density of matter amounts to approximately 10^{-30} g/cm^3.

The scientific community saw the true significance of Hubble's discovery; it was recognized by everyone as convincingly confirming the theory of the expanding Universe. An essential role in establishing relations between astronomy and cosmology was played at that time by the remarkable papers written by A. Eddington and G. Lemaître. They were the first scientists who applied Friedmann's theory to interpret the observational view of the receding galaxies. They saw that the theory provided a clue to the general structure and dynamics of the real Universe. Very soon, in 1931, these authoritative researchers were joined by Einstein, who fully recognized that Friedmann was right. This was not only confirmed by mathematics, but also, one might say, by Nature herself. Later Einstein emphasized in his talks and papers that Friedmann had been the first to step in the right direction in cosmology.

We can guess that it was not easy for Einstein to change his viewpoint. John Wheeler, a well-known American theoretician, said at the Solvay Congress in 1958:

[Einstein] himself tells us of his unhappiness when general relativity predicted that a universe of finite density must have a changing size; of his inventing an artificial new term, with a "cosmological constant," to compensate this "unreasonable" change in size; of the subsequent discovery that the universe really is expanding; and of his conclusion that the cosmological term ought never to have been introduced in the first place.

Having rejected the static model of the world, Einstein at the same time eliminated his universal antigravity as well.

It is good, however, that Friedman solved the problem in a comprehensive way, allowing for the cosmological term: his first investigation, as far as can be judged today, is closest to the real world, and it seems that this world cannot do without antigravity (see below).

Let us go back to galaxies and turn our attention to the Hubble constant, through which relative velocity is linked to relative distance in the expanding world. Hubble himself estimated its numerical value, using astronomical units, as 540 (km/s)/Mpc, where 1 Mpc (megaparsec) = 10^6 pc (*c*. 3×10^{24} cm). But the constant has turned out to be about ten times less. Spectral observations allowed Hubble to find the velocities very accurately, but there was a large systematic error in the distances, which was only eliminated in the late 1950s. According to recent data, the Hubble constant should be within 50 to 65 (km/s)/Mpc. For instance, suppose the distance to a certain galaxy is 20 Mpc; then if $H = 50$ (km/s)/Mpc, we get a velocity of recession equal to 1,000 km/s.

If the Hubble constant is known, there is a possibility of estimating the time from the start of expansion. In the past, the galaxies were evidently closer to each other. What was the time needed for a pair of galaxies to reach a certain distance apart? Let the distance between the two galaxies now be R at a velocity v:

$$t = R/v = H^{-1}.$$

This time, which is obviously the same for any pair of galaxies, is an approximate estimate of the interval during which the recession of the galaxies has been occurring. If $H = 50$ (km/s)/Mpc, it follows that it took 18 billion years from the start of expansion. If $H = 75$ (km/s)/Mpc, it took 13 billion years. The order of magnitude in both cases is about 10 billion years. A time of this order was mentioned, as we recall, by Friedmann, when he gave his "illustrative estimate" of the age of the world.

This estimate of the age of the Universe, which we derived above from a knowledge of the Hubble constant, must nevertheless be regarded as very crude and approximate. It would indeed be adequate if the cosmological expansion could be viewed as a uniform motion proceeding at the same time at all times. In reality, the velocities of the galaxies do not stay constant during the period of expansion. The inherent gravitation of matter tends to curb the expansion, causing the velocities to drop with time. The velocities of expansion are also influenced by the antigravitating vacuum, if it exists; by contrast, the antigravitating vacuum causes an acceleration and tends to speed up the expansion and increase the velocities. Now it must be clear that in order to calculate the world's age accurately, we have to know not only the Hubble constant, but also two more values: the density of matter at present and the density of the vacuum. Since this problem is evidently essential, we shall have to come back to it.

Heritage and tradition

After the late 1920s, Friedmann's cosmology became the subject of in-depth studies. At the time, the most significant problem was to establish and improve its observational basis. The research done by Hubble, by his colleague Milton Humason, and later by other astronomers, added to the stock of concrete facts and observations. Lemaître and Eddington, as well as the Leningrad theoretical physicist Matvei Petrovich Bronshtein, strove to employ these data as fully and consistently as possible for the solution of the fundamental problems posed in Friedmann's papers. The most obvious and urgent problem was that of the real nature of the world's geometry and dynamics. In effect, the development of cosmology during the first three decades of its history occurred mainly within this conceptual framework.

After Friedmann, the first cosmological investigation in the USSR was the research performed by his friend V. K. Frederiks and by A. B. Shekhter. As a university student, Anna Shekhter had been given a topic for her research by Friedmann himself, and she was able to complete it jointly with Vsevolod Frederiks. This was the problem of astronomical measurements in curved space–time, of fixing a frame of reference (a coordinate system) and selecting arbitrary coordinate transformations. The problem is complicated and important; the authors discussed it with their Leningrad colleagues O. D. Khvolson and V. R. Bursian, participants in the seminar where Friedmann had been presenting his reports a while before. The title of the paper was "Calculation of the astronomical aberration and parallax in Einstein's, de Sitter's and Friedmann's worlds from the viewpoint of the general theory of relativity." It was published in 1928 in Russian in the *Journal of the Russian Physico-Chemical Society (Physics Section)* and in German in the same journal that had published Friedmann's cosmological papers.

The following year, in 1929, a paper written by the Pulkovo astronomer A. A. Belopolsky was published. It dealt with an interpretation of the cosmological red shift on the basis of the Doppler effect. Belopolsky was among the first astronomers to employ the effect in order to measure star velocities. He was also interested in various other ways of explaining the red shift in galactic spectra. Can it be that the red shift of light occurs due to "ageing" of the quanta during the millions and billions of years they have been travelling since the moment of their emission? However, this scenario had to be discarded, as it was later shown by M. P. Bronshtein, on the basis of the strict physical theory of light propagation, that physics cannot admit any "ageing" of quanta. The only interpretation that could be substantiated was the one taking into account the Doppler shift and Friedmann's cosmology.

In 1931, Bronshtein published in the journal *Uspekhi Matematicheskikh Nauk* a comprehensive review on the theory of the expanding Universe. Two years later, in 1933, another three reviews of his were published (one of them was co-authored by Lev D. Landau). In 1936, Bronshtein's paper on the quantum theory of gravitational waves appeared. Later it became a classic. This was the first example of a successful synthesis of the two fundamental physical theories of the 20th century: the Theory of Relativity and the quantum theory.[1] The 1936 paper was the first to discuss theoretically the graviton, i.e. the quantum of gravitational waves. The graviton plays the same role in the nature of gravitation as the photon, the quantum of the electromagnetic field, plays in the electromagnetic interaction.

In his 1931 review, Bronshtein wrote that "Friedmann's work was half forgotten" by that time. True, after the first references made by Eddington, Lemaître, Einstein and Tolman (found in Tolman's fundamental book *Relativity, Thermodynamics, and Cosmology*, published in English in 1934 and translated into Russian in 1978), Friedmann's name disappeared almost completely from the pages of articles and books. Starting from the mid 1930s, the role of "the first cosmologist" was played in the West for thirty years by Lemaître, and he was more and more often referred to as "the father of the expanding Universe theory." Hermann Bondi wrote a well-known book entitled *Cosmology* (published in 1952–68 four times in English and becoming a "standard" textbook on cosmology). Friedmann's publications are mentioned in the general list of literature, but there is no reference to their author in the body of the book, although the subject matter of the book is the expanding Universe theory. There is a more recent example. We can read in Jim Peebles's "Physical Cosmology" published in the USA in 1971 (a Russian translation came out in 1975) that the theory of the evolving Universe was predicted by Friedmann. Nevertheless, Peebles believed that Lemaître deserved to be called the father of the expanding Universe theory because Lemaître had been lucky enough to produce his expanding cosmological model when the basic phenomenon, the general recession of the galaxies, was crystallizing, and he had perceived its significance at once.

Lemaître's role in building bridges between theory and observations is

[1] On this paper, and Bronshtein's life and research, see the article (in Russian) by Gorelik & Frenkel listed in the bibliography.

essential and cannot be doubted. It is worthwhile, however, to recall that in his first paper (we shall discuss it below) Lemaître himself indicated Friedmann's fundamental works that offered a comprehensive theoretical solution of the cosmological problem.

Starting from the mid 1930s, the development of cosmology in the Soviet Union took an unfavorable course. Cosmology was steadily attacked with false ideological accusations. Friedmann's legacy was supported by Frederiks and Bronshtein, direct keepers of tradition in cosmology; neither of them was destined to pass the tragic barrier of 1937. In 1938 and later, supporters of the expanding Universe theory were sometimes called "Lemaître's agents."

The initiative in cosmology slipped away from Leningrad to research centers in the West. However, the original enthusiasm in the West caused by Einstein, Friedmann and Hubble faded gradually and gave way to skepticism. At the time, astronomers were using a Hubble constant that was overestimated by an order of magnitude. The world's age turned out to be only 1–2 billion years, which led to an extremely unpleasant contradiction: the age of the Universe proved to be less than the age of the Earth and the Sun (approximately 5 billion years). Fortunately, the contradiction was later resolved by a more accurate Hubble constant, but in those years it was so troublesome that the scientific community distinctly cooled off toward expanding Universe theory. Einstein was definitely losing interest in cosmology. There were attempts to find alternative cosmological theories, and various original interpretations were offered for the red shift. From 1947, cosmology in the USSR became once again the target of criticism from philosophical and political positions.

Speaking at a discussion on G. F. Alexandrov's book *The History of West European Philosophy*, A. A. Zhdanov[2] said:

Today's bourgeois science supplies the church and fideism with new arguments, and these have to be exposed without mercy . . . Without understanding the dialectic progress of cognition and the relationship between absolute truth and relative truth, many of Einstein's supporters . . . come to assert that the world is finite, that it is limited in time and space, and the astronomer Milne even "calculated" that the world had been created two billion years ago.

Philosophers and journalists started exposing cosmology, and they tried to condemn as fiercely as they could the theory which

recently became broadly known in capitalist countries as the expanding Universe theory. The theory was offered in the late 1920s by the Belgian priest Georges Lemaître, and it is underlain by the phenomenon called in astronomy the "red

[2] *Voprosy Filosofii* (Problems of Philosophy), 1947, No. 1, p. 271.

shift." The reactionary scientists Lemaître, Milne and others made use of the "red shift" in order to strengthen religious views on the structure of the Universe ... Falsifiers of science want to revive the fairy tale of the origin of the world from nothing ... Another failure of the "theory" in question consists in the fact that it brings us to the idealistic attitude of assuming the world to be finite.

As we can see, during this well-organized campaign it was quite possible to hand over a major achievement of Soviet science to "reactionary" researchers.

At the artificially inspired discussions on cosmology and cosmogony, scientific arguments were replaced by ideological accusations. But even during the hardest period, cosmological research was continued. The Moscow theoretician A. L. Zelmanov published his first paper in 1938. Later he studied possible generalizations of Friedmann's theory in the direction of a wider range of space symmetries (non-uniform and anisotropic cosmological models). The most important work in the cosmological science of the period was an exhaustive investigation of the expanding world undertaken in 1946 by L. D. Landau's pupil Yevgeny M. Lifshitz, later a Full Member of the USSR Academy of Sciences. His work became a basis for studying the problem of the large-scale structure of the Universe.[3] Very important too was the position taken by Landau with respect to cosmology; together with Lifshitz, he gave an exemplary presentation of Friedmann's cosmology in their famous *Course of Theoretical Physics*. Cosmology was also presented clearly and carefully in V. A. Fock's book *The Theory of Space, Time, and Gravitation*; one may say that Vladimir Fock received Friedmann's theory directly from the hands of its creator. All this paved the way for the new rise of cosmology that began in the early 1960s.

As for philosophical judgements, from the mid 1950s their harsh tone gave way to recognition of Friedmann's theory, first hesitant and vague, but later more decisive. It is true that echoes of forgotten absurdities are sometimes heard even today, when such things are discussed as "finite space," "limited nature of time," and especially "creation of the world from nothing." One needs to be very naive (or to have an ulterior motive) if one sees anything related to "religious superstition and fideism" in these scientific problems. Once again we have to do justice to Friedmann's integrity and courage in setting forth these problems back in the 1920s. Friedmann was setting science clear apart from myths, both ancient and modern.

[3] For more detail see I. D. Novikov, *The Evolution of the Universe*, Nauka, Moscow, 1979 (in Russian) and L. E. Gurevich & A. D. Chernin, *The Magic of Stars and Galaxies*, Mir Publishers, Moscow, 1987 (in English).

Friedmann's follower Lemaître was not only a cosmologist, but also a priest. In 1960–66, he was President of the Vatican Academy of Sciences. At the 1958 Solvay congress, which dealt with cosmology, he said of the expanding Universe theory: "As far as I can see, such a theory remains outside any metaphysical or religious question. It leaves the materialist free to deny any transcendental being." Obviously, Lemaître tends to separate himself from the attitude taken by Pope Pius XII, who in 1951–52 tried to interpret the cosmological inference on the finite age of the Universe in favor of divine creation.

Truly philosophical problems related to cosmology (see e.g. V. L. Ginzburg, *On Physics and Astrophysics*, Nauka, Moscow, 1985) attract wide attention and deserve a comprehensive in-depth analysis.

The following data reveal the dynamics of cosmological research in the Soviet Union. The jubilee volume *Thirty Years of Astronomy in the USSR*,[4] published in 1948, includes a review by Zelmanov of Soviet publications on cosmology until 1947. There are only 46 of them. And only eight deal with the subject matter directly: two of Friedmann's papers, which are presented carefully and praised very highly by the compiler (recall the year), Lifshitz's publication, and other papers that have already been mentioned in this book, including two papers by Zelmanov himself, as well as a paper by M. V. Machinsky "On the finiteness of the world" (1929) and P. K. Kobushkin's paper (1941) on an axisymmetrical model of the Universe. Non-standard explanations of the red shift are attempted in 25 papers, and the rest of them are popular science or philosophy.

Ten years later, the volume *Forty Years of Astronomy in the USSR*[5] did not contain any paper on cosmology, although there is a list of publications. It was supplemented by five papers by Zelmanov. It also included five of Bronshtein's papers dating from 1931–36.

Another decade, and the picture presented in *The Development of Astronomy in the USSR over 50 years*[6] is sharply different: cosmology prospers. A fine review article comprehensively written by the veteran Zelmanov is accompanied by a vast list of papers glamorized by the names of Ya. B. Zeldovich (approximately 50 publications, together with his pupils), V. A. Fock, Ye. M. Lifshitz (with new papers), I. M. Khalatnikov, A. D. Sakharov, M. A. Markov, B. M. Pontecorvo, V. L. Ginzburg, I. S. Shklovsky, N. S. Kardashev ... For the first time in the past three

[4] *Astronomiya v SSSR za 30 Lyet*, OGIZ, Moscow & Leningrad, 1948.
[5] *Astronomiya v SSSR za 40 Lyet*, Fizmatgiz, Moscow, 1960.
[6] *Razvitiye Astronomii v SSSR za 50 Lyet*, Nauka, Moscow, 1967.

decades, there were publications signed by Leningrad researchers; in Friedmann's city the cosmological tradition has been maintained by Prof. L. E. Gurevich.

We cannot give here a complete review of every publication on cosmology since the early 1920s. This is dealt with in a number of fundamental monographs and popular-science books. The most essential monographs are by Ya. B. Zeldovich, *The Structure and Evolution of the Universe*, Nauka, Moscow, 1975 and by S. Weinberg, *Gravitation and Cosmology*, Wiley, New York, 1972. Our goal is much more limited: to outline and explain several studies concerning the issues dealt with in Friedmann's investigations, and to mention briefly modern achievements along the same lines. Moreover, we believe it necessary to refer to the most significant event in cosmology after Friedmann: the prediction and discovery of the relict radiation. This discovery boosted the status of cosmology and changed radically the whole attitude to Friedmann's legacy.

In Newton's language

In 1934, the British theoretical physicists William McCrea and Edward Milne achieved remarkable success in explaining Friedmann's basic results. They did not discover anything new, but revealed the underlying meaning of the cosmological dynamics disguised under the sophisticated mathematics of the general theory of relativity. The theory of cosmology became much clearer. It could now be conceptualized and presented at the level of "conventional" physics.

The British scientists noticed that Friedmann's curvature radius equation, revealing in his theory the dynamics of the Universe, was very similar to the equation representing in Newton's classical theory the motion of a particle in a gravitational field. It is quite possible to start with something uncomplicated and consider the particular case where there is no cosmological term at all (or, as we might say today, the case of vacuum energy density being zero) and there is only "ordinary" matter. Here is what was suggested. Let us assume that matter only fills a sphere of finite radius rather than the whole of space. Now suppose that pressure in the matter is small and can be neglected (pressure was not taken into account in Friedmann's paper either). Suppose the matter within the sphere is distributed uniformly. How will the sphere's radius vary with time?

To solve such a problem in classical mechanics, one must first find out the forces acting upon matter particles in the sphere. Since pressure is disregarded, the whole force is that of gravitation created by all the particles within the sphere. Let us consider a particle on the boundary. It is attracted by all the remaining particles according to Newton's law. This means that the particle is acted upon by each other particle with a force that is proportional to the product of masses of the two particles and inversely proportional to the square of the distance between them. If you add up the forces created by all the particles (i.e. compose the vectors, because it is necessary to take into account the direction of each force), our particle will prove to be acted upon with a force as if all the other particles were at the center of the sphere. (This assertion is valid both in Newton's gravitation theory and in electrostatics, and is called Gauss's theorem.) Therefore, the resulting force must be proportional to the product of the mass of the particle and the total mass of the sphere and inversely proportional to the square of the sphere's radius. Let m be the mass of the particle, M the mass of the sphere and R the radius of the sphere. Then the force (or more accurately its projection on the radius vector) is

$$F = -GmM/R^2.$$

Here G is the gravitational constant. The force is obviously directed towards the center of the sphere: hence the minus sign. The acceleration of the particle under the action of the force equals the force F divided by the particle mass m:

$$a = F/m = -GM/R^2.$$

We can see that the distance of the particle from the center, i.e. the sphere radius, cannot remain constant. Once the particle is acted upon by force, there must be motion, and motion of the particle implies a change in the sphere's radius. Having found the acceleration, we can find the velocity of the particle, and having found the velocity, we can find the distance over which the particle travels, i.e. discover the variation of the sphere's radius with time. That is the technique of solving the Milne–McCrea problem.

The equation of motion in the problem is the equation for acceleration of the particle. It proves to coincide exactly with Friedmann's basic equation reflecting variation of space curvature radius over time in the particular case of a zero cosmological term. (The role of acceleration in Friedmann's theory is played by the second derivative of the curvature radius with respect to time.)

The dynamics of matter expansion outlined by Friedmann's theory can be paralleled very closely by Newton's theory. Naturally, coincidence of the equations cannot be just accidental. This is what reveals a certain correspondence between Newton's mechanics and its generalization, the mechanics of the general theory of relativity. The sphere in the Milne–McCrea problem behaves as if it were "cut out" of the uniform general distribution of expanding matter. The radius of such a sphere varies in direct proportion to the world's radius of curvature.

There is no space curvature in Newton's theory. Therefore, the Milne–McCrea problem stimulated Friedmann's world dynamics rather than the world's geometry. It shows local expansion of matter reflecting the general dynamics of both matter and space itself.

However, what is the source of expansion? The particle on the sphere boundary in the Milne–McCrea model is acted upon by a force attracting it to the center. Therefore, it seems that the sphere would contract rather than expand. The particle under consideration, just like any other particle of the sphere, is acted upon by a force directed towards the center; this is why they start moving and it seems that they cannot do otherwise than fall towards the center. This logic is correct, but it assumes, although this has not been formulated explicitly, a certain "initial" situation: the sphere is first at rest, and then the force is "switched on" and the sphere begins to move. This is a possible particular case (and quite admissible in Friedmann's general solution). It is similar to the fall of a stone in the Earth's gravitational field: the stone is held at some distance from the Earth's surface, then the stone is released and starts falling down, in the direction of the gravitational force.

The same stone, however, can move differently. The stone can be thrown vertically upwards, and then for some time it will move against the gravitational force rather than in the same direction with it. A football can be kicked upwards. A shell can be fired from a gun vertically upwards. A rocket can be fired vertically as well. Having reached a certain topmost point, the stone, football or shell will start falling. But the rocket, if its speed at take-off is great enough, can leave the Earth and never return. The rocket as is well known, has to reach escape velocity to achieve this.

Now we have a sufficient set of straightforward examples to come back to the dynamics of the gravitating sphere and deal with the possible cases of its motion. The first example is that of the fall of a stone; it corresponds to contraction commencing with the state of rest: every particle within the sphere falls freely to the center. This is the case of zero initial velocities. But the velocities at the initial moment can be different from zero. Imagine that in the initial state, at a certain value of the radius, every

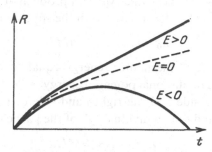

Fig. 4 The radius R of a sphere as a function of time t in the Milne–McCrea model simulates the behavior of the radius of curvature in Friedmann's theory.

particle of the sphere has a velocity that is directed exactly outward, i.e. from the center of the sphere and along a radius. It is clear that at later instants (at least for some short time) all the particles will go on moving in the same direction from the center of the sphere, and the sphere as a whole will keep expanding. The expansion should slow down in the course of time: the motion is against the force of gravitation created by the sphere itself. It is possible to see that if the initial velocities are not very large, the sphere will be unable to resist gravitation for long, so its expansion must sooner or later be greatly slowed down, then the expansion will halt for an instant and then give way to contraction. However, if the velocities are great enough at the start, it is going to be, rather than the case of a ball, the case of a rocket attaining escape velocity. The sphere will expand forever, its particles are going to overcome gravitation, so that their flying apart will never give way to moving back to the center, or contraction.

The uniformity of matter distribution and the Hubble law are closely related to each other in cosmology (see above). In the problem of a gravitating sphere, it is necessary to assume that at the initial instant every particle has a velocity that is proportional, as in the Hubble law, to its distance from the center. Then the uniformity of matter in the sphere will be preserved during its later motion: the particles will not collide with each other, but move away from each other uniformly.

Fig. 4 shows how the radius can depend on time, starting right from zero rather than from an intermediate finite value. This corresponds to expansion from a point, about which, as we recall, Friedmann wrote. But even at this point, every particle has a velocity (directed naturally along the radius from the center).

Qualitative considerations and graphic illustrations can be supported by simple formulae. In the Milne–McCrea problem, it is easy to write the energy conservation law for a particle on the sphere boundary:

$$E_0 = mv^2/2 - GmM/R.$$

Here E_0 is the total mechanical energy equal to the sum of kinetic energy $mv^2/2$ and gravitational potential energy $-GmM/R$. We can divide both the left-hand side and the right-hand side of the equation by the particle mass m, and the "individuality" of the particle will disappear:

$$E = v^2/2 - GM/R.$$

(We have written E instead of E_0/m.)

It is easy to see from the formula that the type of motion, i.e. unlimited expansion or contraction, depends on the sign of the total energy. If the energy is positive, then the sphere's radius can increase indefinitely. This is the case of unlimited expansion. If the energy is negative, the radius cannot exceed a certain maximum value $R_{max} = GM/E$, for which the velocity of expansion becomes zero:

$$v = 0 \text{ for } R = R_{max}.$$

When the radius reaches this value, expansion gives way to contraction.

There is an intermediate case of the dynamics where the energy equals zero. Obviously, unlimited expansion is also possible here, which is shown in Fig. 4 as well.

The law of energy conservation in mechanics follows from the equation of motion (as the result of integrating it). Therefore it is only natural that the formulae in Friedmann's studies include one that is similar to the energy conservation law. Although the general theory of relativity does not contain the notion of gravitational potential energy, a term of the form $-GM/R$ shows up in Friedmann's equations because of the close correspondence between Newton's theory and the general theory of relativity.

If we take another look at the energy conservation law presented above, we can notice that the sign of the total energy, and with it the fate of the cosmological expansion, depends on the relation between the expansion velocity v and the critical value

$$v_c = \sqrt{(2GM/R)},$$

which plays the role of the escape velocity here. Using this value, the energy conservation law can be rewritten as follows:

$$E = (v^2 - v_c^2)/2.$$

It follows that the energy is positive, and therefore the expansion continues endlessly, if the velocity v is greater than the critical velocity (i.e. the escape velocity). Otherwise, the energy is negative and the expansion will eventually give way to contraction.

This form of the conservation law presents a good analogy with our examples (a stone, a ball or a missile) discussed above. From the astronomical point of view, however, it is more convenient to deal with another value, which we shall now consider.

Let us take the Hubble law

$$v = HR$$

and write the expression for kinetic energy (per unit mass) as follows:

$$v^2/2 = H^2R^2/2.$$

In addition, let us take the expression for the potential energy and express the mass in terms of the density and the radius:

$$M = 4\pi\rho R^3/3.$$

Now if the last two equations are substituted into the energy conservation law and both sides of the equation are divided by $4\pi GR^2/3$, we obtain

$$3E/4\pi GR^2 = \rho_c - \rho.$$

Here appears a new value

$$\rho_c = (3/8\pi G)H^2,$$

which is called the critical density.

It can now be seen that the sign of the total energy depends on the difference between the critical density and the matter density: if the density is less than the critical value, the energy is positive; if greater, the energy is negative; if the density equals the critical value, the energy is zero. Here then is a chance to solve the problem as to which of the three above-mentioned dynamic possibilities is true in the real world. This would constitute an answer to the problem of what kind of future lies in store for the Universe. The procedure is as follows: using the Hubble constant, it is necessary to find the critical density and then to compare the result with astronomical data on the average density of matter in the Universe. If the Hubble constant is within 50 to 75 (km/s)/Mpc, the critical density is in the range

$$\rho_c = (0.5\text{--}1.0) \times 10^{-29} \text{ g/cm}^3.$$

What is known on the matter density in the Universe? In the early 1930s, Hubble estimated the density of matter in the world to be about 10^{-30} g/cm^3. The latest methods of counting the number of galaxies yield a lower average density of luminescent matter, which is approximately three to six percent of the critical density. Both Hubble and modern calculations estimate the matter density as less than critical. If this is so, the total energy in the energy conservation law is positive. The conclusion should follow that the Universe expands indefinitely. But the real situation is much more complicated. There are two essential points: the presence of "hidden" masses, and the action of the antigravitating vacuum on the matter in the Universe.

Besides the luminescent matter of the galaxies, the Universe is certain to include some dark, invisible matter whose average density may turn out to be significantly greater than the density associated with the galaxies. Back in the 1930s, the Swiss astronomer Fritz Zwicky found out that the total gravitational mass of compact clusters of galaxies (including hundreds and even thousands of members) is noticeably greater than the sum of the masses of the constituent galaxies. It was suggested that the deficient mass was probably that of the intergalactic gas which could not be observed at the time. During the 1970s, the intergalactic gas was detected by its X-ray radiation (i.e. by the electromagnetic waves of the X-ray range it emitted). There are considerable quantities of the gas, and its total mass is often comparable to the total mass of the galaxies in clusters. However, it is still five to ten times less than is needed to explain the observed total mass of the clusters.

It is worthwhile to mention briefly how the total mass is determined. The galaxies move within a cluster at tremendous velocities reaching thousands of kilometers per second (the velocities can be measured by observing the spectra). However, the force of gravitation created by the cluster as a whole does not allow galaxies to leave the cluster. The galaxies perform some limited, finite motions within the cluster. In other words, they move at velocities not exceeding the "escape velocity" for that particular cluster. If this is so, we can take the size of the cluster and typical velocities of the galaxies and attempt to make an estimate of the total mass producing the force of gravitation. The escape velocity can be found using a formula of the same form as for the gravitating sphere: $v_c \approx \sqrt{(GM/R)}$, where M and R are now the mass and the radius of the cluster. If v, the velocity of the galaxies, is less that this velocity or, at least, is comparable to it, the cluster mass can be estimated as follows:

$$M \gtrsim v^2 R/G.$$

Substituting here a typical velocity $v \approx 10^8$ cm/s and a typical cluster radius $R \approx 10^{25}$ cm, we obtain $M \gtrsim 10^{48}$ g, i.e. 5×10^{14} solar masses.

The mass that has been estimated in this way includes the matter within the volume of the cluster, i.e. the galaxies, the intergalactic gas and, which is the main point, the hidden matter that does not emit any light and reveals itself by its gravitation only. And the gravitation of the cluster is produced basically by this hidden dark matter. Moreover, according to the data obtained by Jaan Einasto from the Tartu Observatory (and the data are supported by other astronomers), the hidden mass of the whole set of clusters (and of smaller associations of galaxies) makes the greatest contribution into the average density of matter in the Universe.[7] Taking into account the hidden mass, the average density amounts to approximately 10^{-29} g/cm^3, i.e. it seems to be comparable to the critical density: it may even be somewhat greater.

It is clearly essential to find out what that hidden matter is. The problem of the nature of hidden masses is very challenging. The following candidates have been suggested for the role of "carrier" of the invisible gravitating matter: stars of very small mass and very low luminosity (a name has already been devised for such hypothetical stars: black dwarfs, and those are stars and not for instance, black holes); planet-like bodies like Jupiter, which do not emit light themselves; elementary particles that are regarded as possessing some rest mass and interacting between themselves and other matter, for all practical purposes, through gravitation only: these can be either known particles, such as neutrinos (if they do possess a rest mass), or hypothetical particles that have not yet been observed experimentally although modern concepts of elementary particle physics give grounds to discuss both the possibility of their existence and their properties. We shall return to the problem at the end of the chapter.

The cosmology of the antigravitating vacuum presents yet more problems than the hidden masses. It can be said to be hidden from us even further. If it does exist (and we shall see very soon that it seems to do so), the vacuum is capable of exerting influence on the dynamics of expansion, and the inference on the future of the Universe depends immensely on the

---

[7] See e.g. J. Einasto, A. Chernin & M. Jover, "Hidden mass in galaxies," *Priroda*, No. 5, 1975, pp. 39–43.

presence of the vacuum. And this is not all. The role of the vacuum in Friedmann's cosmology deserves special consideration.

Vacuum and expansion

Friedmann's general solution for spaces of positive and negative curvature contains, as we have seen, a cosmological term whose presence means consideration of the antigravitating vacuum (recall its modern interpretation first mentioned above in Chapter 9).

The antigravitating vacuum is a very specific state of matter that was not yet known to Newton's classical mechanics. Nevertheless, its dynamic role in cosmology can be discussed in Newton's terms by extending McCrea and Milne's technique to cover this case. It is necessary to suppose that the expansion of the sphere which we imagined to be separated from the general distribution of matter occurs in a vacuum that fills the whole world with a constant, time-independent density. Then the particles of the sphere will be acted upon by the spherical volume of the vacuum within it.

It is necessary, however, to take into account one more point. According to the general theory of relativity, not only is matter density a source of gravitation, but matter pressure is a source of gravitation as well. If the medium is uniform, matter density ρ and pressure p yield the "effective gravitational density"

$$\rho_{ef} = \rho + 3p/c^2.$$

The density ρ_{ef} is effective in the sense that this density should be employed to calculate the mass for Newton's law of gravitation. For the sphere under discussion, this mass is given by

$$M_{ef} = \rho_{ef}(4\pi R^3/3).$$

As we indicated above, the vacuum corresponds to a pressure that is equal in magnitude, although the sign is opposite, to its energy density:

$$p_V = -\epsilon_V = -\rho_V c^2.$$

It is clear now that the effective gravitational density for the vacuum, given by

$$\rho_{efV} = \rho_V + 3p_V/c^2 = -2\rho_V,$$

is negative if the vacuum density is positive. This is how the antigravity of the vacuum appears: if the effective density is negative, the corresponding gravitational mass M_{efV} is negative as well, and therefore this mass will repel particles rather than attract them.

It is easy now to infer the equation of motion for a particle on the boundary of the sphere. The particle will move with an acceleration of

$$a = F/m = - GM_{ef}/R^2.$$

Here the effective gravitational mass is understood as a value taking into account both matter and vacuum:

$$M_{ef} = (\rho - 2\rho_V)(4\pi R^3/3).$$

The pressure in matter is neglected, and the pressure in the vacuum creates a negative term in the expression for the effective density and mass.

The equation of motion (i.e. the equation for acceleration) underlies the mechanical energy conservation law:

$$E = v^2/2 - GM/R + GM_V/R.$$

Here (unlike what we have seen above) the potential energy is represented by two terms: one is related to matter itself and the other to the vacuum. The vacuum mass M_V is $\rho(4\pi R^3/3)$, included in a sphere with radius R.

Friedmann's papers contain the equation we have just presented. (By velocity v is meant the derivative of the radius R of curvature with respect to time, so this is a differential equation needed to find the radius as a function of time.) The first cosmological paper by Friedmann deals with this equation under various conditions relating to the vacuum density. And this is precisely where Friedmann's conclusions on the nature of expansion follow. (While performing the analysis, the possibility of either a positive or a negative value of vacuum density was admitted. Recall that the vacuum density is a constant value that depends on neither time nor coordinates.)

As a particular case, this equation allows the possibility of a static solution. If the solution is static, both the acceleration and the velocity are zero. Therefore we can take the acceleration equation and find the relation between the densities of matter and vacuum:

$$\rho = 2\rho_V.$$

This is the same relation that Einstein had in his static model of the world. The gravitation of matter in the static solution is exactly balanced by the antigravitation of vacuum. Hence the possibility of the state of rest.

And what if the exact balance between gravitation and antigravitation is tipped? For instance, suppose for some reason the matter density becomes less than the vacuum density multiplied by two. What then?

Obviously, the sphere must start expanding, because the antigravitation of the vacuum must win over the gravitation of matter. The antigravitational effect of the vacuum will "push" the sphere particles in the outward direction from the center, and the expansion will occur with acceleration (and this acceleration will increase). But the matter density can only drop because of the expansion, therefore the lost balance of gravitation and antigravitation will never be regained. Matter will expand indefinitely.

This type of solution within Friedmann's general picture was thoroughly studied by Lemaître in 1927. Remarkably, this was still two years before Hubble published his paper on the red shift of the galaxies, and Lemaître gave the following title to his paper: "A homogeneous universe of constant mass and increasing radius accounting for the radial velocity of extra-galactic nebulae." Like Eddington, he correctly interpreted the prevalence of red shifts in the available observational material and came to the conclusion that "the receding velocities of extra-galactic nebulae are a cosmical effect of the expansion of the universe."

In his theoretical analysis Lemaître proceeded, as we can well see, from a rather particular, special version of the dynamics of the Universe. Besides, he assumed that a three-dimensional space had a positive curvature, just as in Einstein's world. However, in contrast to Einstein's static model, in Lemaître's model expansion occurs, because the gravitation of matter is not balanced but overcompensated by the antigravitation of the

Georges Lemaître.

vacuum. The expansion can start from the state of rest, just as in Einstein's theory, if for any reason the balance of gravitation and antigravitation in it is disturbed.

It is worth mentioning that Lemaître made a special remark, considering it necessary to refer to the founder of evolutionary cosmology. Lemaître points out that the problem of the non-static Universe was earlier given full mathematical treatment by Friedmann, although it was unrelated to observation of nebulae.

Eddington, being interested in Lemaître's paper, translated it into English and submitted it for publication in the *Monthly Notices of the Royal Astronomical Society*, where the paper appeared in 1931 and aroused wide interest. The interest was quite justified: that was the first attempt to apply the new cosmological theory to the real world on the basis of new data obtained by observation. Lemaître preferred a particular version of Friedmann's solution that retained a link with Einstein's static model, but the main point was expansion. It is called the Lemaître–Eddington model.

Lemaître went beyond the original model and in the 1930s–40s continued to study the expanding Universe theory, striving to make it as close as possible to real astronomical observations. He was clearly aware of the fundamental problem in cosmology posed by Hubble's discovery: we can see that there is a general expansion of the Universe, but according to which law and in which geometry? Lemaître hoped that observations would allow Friedmann's theory to be given the concrete form that its founder could only dream of.

Indeed, Friedmann's theory allows various scenarios of expansion dynamics, which differ primarily in the duration of expansion: the expansion is either unlimited in terms of time or gives way to contraction. Also the theory includes several possible space geometries; although space is uniform and isotropic, it is curved and the sign of the curvature is not known in advance. As we recall, Friedmann gave his solution of the "world equations" for three-dimensional spaces of either positive or negative curvature.

It is worth mentioning here that there is a third possibility for a three-dimensional space geometry: its curvature may be equal to zero, in which case it is Euclidean. Einstein and de Sitter analyzed the possibility and extended Friedmann's theory in 1932; that was their contribution to the expanding Universe theory.

But how difficult is each new step: ten years had to pass after Friedmann's first paper, and two outstanding intellects, those of Einstein and

de Sitter, had to combine, just in order to supplement Friedmann's theory
with a variant assuming the most elementary metrics. (Note that the
four-dimensional space–time in the three-dimensional Euclidean space is
curved due to the presence of gravitating matter in it.)

Now what is the real dynamics and geometry of the one Universe in
which we are fated, as Friedmann put it, to live?

About seven decades have passed since the prediction and discovery of
the expansion of the Universe. During this period, cosmology has
changed greatly, it has been enriched with new ideas, first-class theoretical
results and astronomical discoveries, but there still seems to be a long way
to go before the problem is finally and reliably solved. There were periods
when the problem appeared to be solved or the solution was thought to be
just around the corner. For instance, after the early 1960s many researchers believed the situation to be clear. Firstly, the frustrating contradiction
as to the age of the Universe had been removed at last. Secondly,
adequately substantiated estimates of the average density of luminous
matter had been produced. They gave a density that was appreciably less
than the critical density, and the conclusion was drawn that expansion
occurs with a positive total energy (using Newton's terms), and will
therefore continue for an unlimited period. The vacuum and its density
were not considered then, and hidden masses were not taken into account.

Under the circumstances, the picture might really seem to be simple
enough. According to Friedmann's theory, there is a mutual unam-
biguous relation between the dynamics of the Universe and its geometry.
In Newton's terms, if the total energy of expansion is positive, then the
curvature is negative; if the energy is negative, the curvature is positive;
and if the energy is zero, the curvature is zero as well. This is true,
however, only if there is no antigravitating vacuum. On the assumption
that there is no such vacuum, it was asserted during the 1960s that the
data on the low density of matter implied not only irreversible expansion,
but also negative space curvature. This negative curvature was commonly
associated with an infinite volume of the Universe, an arbitrary assumption thus being made (at least tacitly) about the general topology of space.
In fact, Friedmann's warnings on this point had been forgotten. Specialists and lecturers, not to mention popular-science writers and journalists,
were quite confident that the Universe could expect an endless future, and
its overall size was also thought to be arbitrarily large.

Clouds started to appear on the horizon in the mid-1970s. J. Einasto
from the Tartu Observatory (Estonia) and J. Ostriker and J. Peebles from
Princeton (USA) point out that the old problem of hidden masses should

be taken seriously. Step by step, it became clear that the hidden masses may play a key role in cosmology: on them depends the estimate of the density of matter in the world, with all its consequences (see above). And one more consideration emerged later, during the mid-1980s, and made researchers take a fresh look at the cosmological vacuum as well.

The point is that if there exist considerable hidden masses, bringing the general density of matter to the critical density, the age of the world produced by the theory should be less. Accurate calculations on the basis of Friedmann's theory (by the way, there is an explicit formula for the world's age in Friedmann's first paper) reveal that by taking the density to be critical and the Hubble constant to be within 50 to 75 (km/s)/Mpc, we can arrive at an age of 8.7 to 12 billion years. This estimate is also approximately 10 billion years, but not 13 to 18 billion as in the low-density world (where the age is very close to the reciprocal of the Hubble constant).

On top of that, a few years ago some astronomers revised the earlier estimates and came to the conclusion that the age of the oldest stars in our Galaxy had to be no less than 14 billion years, if not 17 billion. Moreover, new data on the processes involved in the synthesis of heavy elements in the stars and the data on the diffusion of certain radioactive nuclei may result in a still greater estimate for their age, up to 22 billion years. It is clear that the age of the world cannot in any case be less than the age of the oldest stars or nuclei. But this means that cosmology runs into another direct contradiction related to the world's age.

The most convincing considerations on the problem are based on the return of the antigravitating vacuum into cosmology. Taking Friedmann's general formula for the age of the world and demanding that the formula should yield not less than a billion years (this is a very modest requirement), we shall soon see that the vacuum density should not only be non-zero, but also its value should exceed the density of matter. Detailed calculations recently carried out on the basis of the formula by the German researchers H. Klapdor and K. Grotz indicate that the vacuum density should be no less than five to ten times the density of the luminous matter of the galaxies, if the Hubble constant is 50 (km/s)/Mpc. If the Hubble constant is taken to be 75 (km/s)/Mpc, then the vacuum density turns to be even greater: it exceeds the density of the luminous matter by a factor of 15 to 30 times. Moreover, the vacuum density should exceed the total matter density, including the hidden masses. With a Hubble constant of 50 (km/s)/Mpc, the excess should be estimated at 20 to 30 percent, and at 75 (km/s)/Mpc the excess should be much greater, by

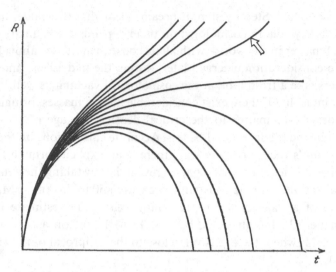

Fig. 5　Various ways in which the radius of curvature can behave, taking antigravity into account. This figure reproduces the graphs first drawn by Lemaître and recently found in his archive by the Polish cosmologist M. Heller. The arrow indicates the type of behavior which is regarded (as of today) as the most probable.

a factor of four to eight. It follows that the vacuum, since it possesses significant density, generates a repulsion so great as to be stronger than the gravitation of matter on the whole throughout the Universe. Under these circumstances, cosmological expansion should go on without restriction in terms of time (Fig. 5).

The same calculations bring about a conclusion on the sign of three-dimensional space curvature: the curvature is positive, as in Friedmann's first paper, if the density of the hidden masses is not less than twice the density of luminous matter. According to what has already been said about hidden masses, this condition should be regarded as true.

Therefore, today the world picture in cosmology is as follows: the real Universe expands with no restriction in terms of time; the three-dimensional space where the expansion occurs has a positive curvature. Antigravitating vacuum prevails in the Universe. The time elapsed since the start of expansion is at least 17 billion years.

Let us wait and see how long this picture will hold. Can it be the ultimate solution of the problem?

How striking is the destiny of Einstein's antigravity! It was first invented for reasons that later turned out to be incorrect, then it was rejected by the inventor himself as the greatest scientific mistake he has ever made (in his own words), and now antigravity appears once more, this time to save the cosmological theory.

However, even before now interest in this idea of Einstein's has occasionally been revived. We can start from the fact that in Friedmann's theory the antigravitating vacuum, in the form of the cosmological term, is present from the very beginning and fully taken into account. It is also worth mentioning that about 30 years ago the Leningrad theoretical physicist E. B. Gliner was the first to point out that it was precisely vacuum rather than any other state of matter. On the basis of the general theory of relativity, he proved that one cannot distinguish between rest and motion with respect to such a "medium" in which energy density equals pressure with the sign reversed. And this is the first and principal property of vacuum. Later, in the early 1970s, the Moscow physicist A. A. Starobinsky found some convincing arguments supporting the profound relations between the antigravitating vacuum and the vacuum states studied in the physics of the microcosm. Then E. B. Gliner, I. G. Dymnikova, a theoretical physicist at the Ioffe Physico-Technical Institute, and L. E. Gurevich, Professor at the same institute, suggested and developed a concept of the antigravitating vacuum as the cause of cosmological expansion.

They posed the following problem: if antigravity is capable of repelling physical bodies from each other, can it be that it was antigravity that gave every particle in the Universe the initial push which is observed even now in the recession of galaxies? To make it possible, we should assume that the vacuum density at the start of the history of Universe was enormous, it exceeded by many orders of magnitude the value cited above for the modern state of the world. It is better to assume that at the start there was no matter at all, as it were, and there was only vacuum. For reasons that are not yet exactly known, at some point in time the energy stored in the vacuum generated matter, and the matter could not avoid expanding under the antigravity of the vacuum. From this viewpoint the initial state of the Universe should be assumed to be the state identified in de Sitter's cosmological theory: de Sitter suggested that there is no matter, there is only the cosmological term, i.e. vacuum. (In effect, this is also a particular case of Friedmann's theory.)

Naturally, de Sitter himself did not employ the terms of "antigravitating vacuum." However, he stated clearly enough that two particles in his world would immediately start moving apart. Eddington also pointed out this property of de Sitter's world, explicitly relating it to the observed recession of galaxies. At present, de Sitter's world and the processes occurring in it are studied by researchers in many countries; they can see in it a real prototype of the initial, "pre-Friedmann" state of the Universe. Especially significant are the results achieved by Alan Guth in the USA and A. D. Linde in the USSR. They have succeeded in connecting cosmological ideas with the latest physics of elementary particles. The theory they advance is called the inflation model, or the model of the inflating Universe.[8] The point is that the vacuum produces very fast expansion, as if "blowing bubbles" and their number can even be infinite. They can make up a "space–time foam." One of the bubbles became our Universe.

This exciting picture is being developed now. There is a lot of room in it for fantasy and intriguing findings. Some of it has been given theoretical backing; the rest is still hypothetical. Some challenging problems can be seen. For instance, how can it happen that after the inflation the vacuum density drops abruptly by many orders of magnitude and exactly to the value that follows from the current data ($c.10^{-29}$ g/cm^3, see above)? There is much work in store here, and so far the inflation theory is very far from being on a par with the theory of expansion or the "hot" Universe theory.

George A. Gamow.

[8] See the books *The Past and the Future of the Universe*, Nauka, Moscow, 1986; I. L. Rozental, *Geometry, Dynamics, and the Universe*, Nauka, Moscow, 1987.

A hot universe

There was a period when physicists and especially cosmologists argued a lot about the forthcoming "heat death" of the Universe. Much more fruitful, however, proved to be a discussion of the "heat birth" of the Universe. And indeed, what can be said on the thermodynamics of the early Universe, on the temperature of matter at the very start of expansion? Was the Universe hot or cool during the epoch when it consisted entirely of the pre-star, pre-galaxy original cosmic medium? These were the questions posed by G. A. Gamow in the late 1940s; he had started his cosmological studies back in 1924 at Leningrad University under the guidance of Professor A. Friedmann.[9]

As has already been mentioned, the theory of the expanding Universe ran into a lot of trouble in the 1940s–50s because of a too low estimate of the world's age. The alternative non-Friedmannian cosmology of the "steady-state theory" emerged and spread in the West as a response to this situation. The theory was enthusiastically developed at that time by H. Bondi, T. Gold and F. Hoyle. According to the steady-state theory, the Universe as a whole has always been just as it is today. The galaxies do recede, but the process of the world's general expansion is accompanied by continuous creation of the new matter, which fills the growing gaps between the stellar systems and becomes building material for new galaxies. If this is so, the problem of the early history of the Universe does not exist at all, because there was no such thing as the early Universe.

These ideas are not without wit, and they cannot be rejected point-blank: they do not oppose astronomical observations. The suggested rate of creation of new matter is so small that from the viewpoint of physics the theory does not seem to be a very bold one, seeing that it cannot in any case be settled in the course of direct laboratory experiments. At the same time, there has never been a chance to prove that matter is born continually. Many physicists believed then that cosmology had reached an impasse: it was so easy in it to produce abstract and even fantastic ideas that could be neither confirmed nor rejected.

Any appeal to evolutionary cosmology was most often greeted skeptically by the scientific community. It was rather a subject for jokes and irony than a subject for in-depth studies. Steven Weinberg, an author of the theory of the electroweak interaction, wrote: "I remember that . . . in

---

[9] Georgy Antonovich Gamow (1904–68) worked at the Physico-Technical Institute in Leningrad in 1931–33 and in the USA starting from 1934. Apart from his studies in cosmology, he gave the first quantum-mechanical explanation of gamma-ray decay and was the first researcher to produce calculations of the genetic code.

the 1950s, the study of the early universe was widely regarded as not the sort of thing to which a respectable scientist would devote his time."[10]

Naturally, for a physicist of Gamow's stature, general opinion did not matter much. Besides, his early conviction that the theory created by his teacher was valid appeared to be so deep that he did not attach too much significance to the discrepancies in estimates of the world's age. The contradiction was only resolved in favor of Friedmann's theory at the start of the 1960s. Gamow's work had begun long before that, and it opened up amazingly broad vistas, as it turned out later. Gamow predicted remarkable discoveries in astronomy, which were later actually made. Friedmann's theory was finally proved and confirmed in the process. Gamow was able to reinstate cosmology in its proper place in science, and he succeeded in winning worldwide recognition for his expanding Universe theory.

Before discussing Gamow's principal cosmological paper, let us mention briefly his earlier works, whose purpose was to relate cosmology and cosmogony. If the Universe was really born in a state of very high density, it is clear that the stars and galaxies we can now see could not have existed forever but must have appeared at a certain stage of the cosmological expansion. At the beginning, all the matter of the Universe was distributed more or less uniformly over the space of the world, and then the matter started to segregate into separate condensations which eventually turned into galaxies and stars. Gamow suggested that the physical mechanism underlying such fragmentation of the space medium was the gravitational instability first mentioned by Newton. Jointly with Edward Teller, a well-known nuclear physicist, he carried out in 1939 a theoretical investigation to find out how the forces of gravitation acting between all the particles in the Universe could manifest themselves in the general cosmological expansion. These forces proved to be able to curb the expansion in the areas of the medium where its density for some reason was at least a little greater than the average. To put it more accurately, the pull in such areas is a bit stronger than within the whole medium, and therefore denser areas should expand a little slower than the entire medium. This is why the excess of their density over the average will grow with time. Sooner or later, the expansion of these areas will be blocked completely by their own gravitation, and this will give rise to the

---

[10] From his well-known book on the cosmology of the early Universe, *The First Three Minutes*, Basic Books, New York, 1977, p. 4. Mention should also be made of his excellent book *Gravitation and Cosmology*, in which the role of Friedmann's classic papers is correctly described and evaluated.

The Kharkov Conference, 1929. The photograph shows a number of people mentioned in this book. First row, left to right: unknown, A. I. Tudorovsky, V. R. Bursian, V. A. Fock, D. D. Ivanenko, Ya. I. Frenkel, W. Heitler, P. Jordan, J. Grommer. Second row, above Ivanenko and Frenkel respectively: L. D. Landau and G. A. Gamow. Third row: between Landau and Gamow, A. V. Timoreva; left of her, I. Ye. Tamm, A. O. Gelfond, L. V. Rozenkevich; right of her, A. N. Arsenieva-Geil, V. K. Frederiks. Back row: ninth from left, V. A. Ambartsumian; eleventh, Yu. A. Krutkov.

emergence of linked and non-expanding condensations, i.e. galaxies and stars.

Even today this pattern serves as the basis scenario in the cosmogony of galaxies and in modern theories of the evolution of the large-scale structure of the Universe. Gamow and Teller only indicated a qualitative mechanism of gravitational instability. Its detailed general quantitative theory was produced on the basis of Friedmann's cosmology by Ye. M. Lifshitz in 1946 (the theory has already been mentioned above). Later, in 1970–80, the theory was substantially extended and developed (with due regard to non-linear processes) in the work of Academician Ya. B. Zeldovich and his colleagues. Simultaneously, success was achieved in this direction by R. Dicke, J. Peebles, J. Silk, J. Ostriker, M. Rees and other major physicists in the West.

Another trend in research on cosmogony that drew Gamow's attention was related to the amazing fact that practically all galaxies in the Universe rotate. Fast-rotating galaxies commonly reveal a specific spiral pattern by which they are recognized. Our Galaxy and the Andromeda Nebula are well-known to be spiral. Gamow assumed that rotation of galaxies was due to an intense vortical motion that must have been present in the pre-galactic medium just before its fragmentation. He assumed that there were pre-galactic vortices in the Universe from the very start. He wrote that both the general cosmological expansion and the local rotations imposed on it, i.e. the vortical motions, are of the same nature and were generated simultaneously by the "big bang" that produced the Universe. Later research showed that such primary vortices could not exist (in particular, they would be inconsistent with the high isotropy of the relict radiation – see below). The reason for the pre-galactic vortices should have been looked for in the hydrodynamic processes accompanying gravitational instability. The instability should produce rather fast motion of large volumes and masses of the medium but does not generate vortices directly. It was found out that vortices can and indeed must appear by the general laws of hydrodynamics, in gravitationally induced motions, even if there are no vortices at the start. It is noteworthy that Friedmann was very much interested in one of the processes bringing about vortices in a flow (this was mentioned in Chapter 7), but it was related to atmospheric vortices rather than to galaxies. Atmospheric conditions are different, but the physics of the processes prove to be the same everywhere. The development of vortices in cosmological conditions was first studied theoretically in the 1970s at the Ioffe Physico-Technical Institute. Therefore, Friedmann's vortices were transferred, as it were, from meteorology into Friedmann's cosmology and proved to be able to impart fast enough rotation to spiral galaxies.[11]

If we reverse the cosmological expansion in our imagination and consider the world's evolution backwards in time, then the Universe will seem to be contracting. All bodies become warmer if they are compressed and cooler if they expand: something like that is probably included in any elementary textbook on physics. Most often this is true. In 1948 Gamow suggested that the Universe as a whole obeys this "school" rule as well. This means that the Universe had to be very hot at the start of its existence.

Gamow's history of the early Universe can be presented (considering some of the points that were refined later) as follows. First the Universe was created, and little detail can be given so far concerning this event. By contrast, the history of the Universe from a very young age, e.g. one hundredth of a second after its birth, can be studied reliably, on the basis of general physical laws. The density of matter at that moment amounted to about 10 billion grams per cubic centimeter. This density is obviously enormously large, but it is still tens of thousands of times less than the density of the most compact stars, i.e. pulsars, that are well-known today both in theory and from direct astronomical observation.

As for the temperature, it was 100 billion degrees when the Universe was one hundredth of a second old. When matter is so hot, neither molecules nor atoms nor even atomic nuculei can exist: if they had existed, they would have been destroyed and broken into separate component particles because of the very fast thermal motion. The matter of the Universe consisted then of free elementary particles. It was a mixture of the lightest particles, such as electrons and their antiparticles positrons, photons, neutrinos and antineutrinos, represented in approximately equal quantities; there were also very small quantities of heavy particles, i.e. protons and neutrons, about a billion times less than the quantities of electrons per unit volume.

Because of expansion, the Universe was cooling, and three seconds after its birth, when the density reached that of water, the temperature of matter amounted to a billion degrees. That was now "cool" enough for thermal motion not to prevent protons and neutrons from joining each other and producing atomic nuclei. At first only the simplest nuclei appeared: those of deuterium, a heavier variety of hydrogen containing a

---

[11] More detail can be found in Russian in "Hydrodynamics of the Universe" by A. D. Chernin, *Priroda*, 1976, No. 6, pp. 106–115 and in English in *The Magic of Galaxies and Stars* by L. E. Gurevich and A. D. Chernin, Mir Publishers, Moscow, 1987.

proton and a neutron. A deuterium nucleus might then encounter another proton or another neutron and produce, respectively, a nucleus of helium-3 (two protons and a neutron) or a nucleus of tritium, the heaviest hydrogen isotope (a proton and two neutrons). A helium-3 nucleus might then collide with a neutron, and a tritium nucleus might collide with a proton. In both cases, a nucleus of "ordinary" helium-4 is produced (two protons and two neutrons). But heavier nuclei than these could not in practice be formed: any further collisions between particles were prevented by cosmological expansion. By the end of the first four minutes, there was a specific chemical composition in the Universe: approximately 70–75 percent of the mass consisted of hydrogen nuclei and approximately 25–30 percent of helium-4 nuclei. Apart from the nuclei, there were photons, neutrinos, antineutrinos, as well as electrons, the number of electrons now being equal to that of protons (including those in the nuclei). There were no positrons anymore: they had been annihilated somewhat earlier, i.e. each of them encountered an electron and collided with it, and the pair disappeared, imparting their energy to photons, neutrinos and antineutrinos.

After the epoch of nucleus formation, nothing particular occurred in the expanding Universe for almost one million years. By the end of that period, the temperature of matter had fallen to several thousand degrees, which was "cool" enough for electrons to be able to join the nuclei and yield atoms of hydrogen and helium. From this material there would later be formed the first stars and galaxies.

Now what about photons, neutrinos and antineutrinos? They could survive happily despite all changes and transformations in the matter, and they exist in the Universe now, their number still being the same. But their concentration and temperature are naturally much less now than in the early Universe. These particles are now remnants, leftovers from the early hot state of the Universe. They are termed, at I. S. Shklovsky's suggestion, relict particles.

This is the prehistory of the Universe restored by the power of theory. In its basic, fundamental points, the prehistory was revealed by Gamow, who worked with his pupils Ralph Alpher and Robert Herman; the famous physicist Enrico Fermi helped them in their calculations of nuclear transformations. They succeeded in developing the thermodynamics of the Universe when they described and explained its thermal evolution on the basis of the general principles of statistical physics. They also explained the chemical composition of the matter from which the first astronomical bodies had been formed. It was shown that the observed cosmic average of the mass ratio between hydrogen and helium (four to one or, probably, three to one) could be brought about during the very first minutes of the Universe. That was one of Gamow's original aims, and it was achieved because nuclear physics had been introduced into cosmology.

Here it is worth quoting an authoritative witness of the time, Steven Weinberg:

It was extraordinarily difficult for physicists to take seriously *any* theory of the early universe. (I speak here in part from recollections of my own attitude before 1965.) ... The first three minutes are so remote from us in time, the conditions of temperature and density are so unfamiliar, that we feel uncomfortable in applying our ordinary theories of statistical mechanics and nuclear physics ... Even worse, there often sems to be a general agreement that certain phenomena are just not fit subjects for respectable theoretical and experimental effort. Gamow, Alpher, and Herman deserve tremendous credit above all for being willing to take the early universe seriously, for working out what known physical laws have to say about the first three minutes. *The First Three Minutes*, pp. 131–2.

Apart from what Gamow wanted to achieve in his theory, he received a "by-product," or reward for his scientific courage. This was the relict particles. Theory predicted their incontrovertible existence in the modern Universe and even allowed Gamow to find their approximate temperature. The temperature can be calculated from the need to match the observed ratio between the cosmic abundances of hydrogen and helium (and deuterium as well). The papers Gamow and his co-workers published indicate the result of such calculations: the temperature of the relict particles, primarily photons, must be very low at present, close to absolute zero: only a few degrees Kelvin.

The relict photons were discovered directly and registered in 1965. Their temperature proved to be approximately three degrees Kelvin – in magnificent agreement with the theoretical prediction.

In terms of waves rather than particles, the three-Kelvin photons should be matched by electromagnetic radiation in the range of short, microwave and radio waves (wavelengths from 0.01 to 10 cm). Such waves propagate in the Universe in all directions, and the relict radiation fills its entire volume, thus creating an all-pervading background radiation.

Specialists in radio astronomy believe that the technology needed to register this radiation has existed, in principle, since the late 1940s. But Gamow's colleagues talked to experts from the USA National Bureau of Standards, who gave them a negative answer. In 1965 the relict radiation turned up, so to speak, on the doorstep of the Pulkovo Observatory: a horn antenna there, designed by S. E. Khaikin and T. A. Shmaonov, proved to be sensitive enough to detect it. But nobody thought is mattered.

The 1965 discovery was made quite accidentally. The American astronomers Arno Penzias and Robert Wilson[12] had heard nothing about microwave relict background radiation. They were trying to adjust their antenna for other purposes. It was tuned to the 7.35 cm wavelength, and they found that this was associated with unavoidable "noise," in fact a signal that arrived from every direction with equal intensity. As soon as the Princeton physicists heard about it, they recognized the received background radiation and interpreted it as a relic of the early Universe that could have been expected according to Gamow's "hot" theory. This finding can well be compared to Hubble's discovery.

Thus the reality of the early Universe was made obvious. Friedmann's evolutionary cosmology was supported clearly and convincingly.

During the past quarter of a century or so, the relict radiation has been studied thoroughly. Observations have been carried out at various wavelengths. The temperature so defined amounts to 2.8 K (Kelvin) according to the most accurate modern measurements. The relict background has revealed a high degree of uniformity and isotropy.[13] The most important results were obtained by a group headed by Yu. N. Pariysky, Corres-

Arno Penzias and Robert Wilson.

[12] In 1978 they were awarded the Nobel prize in physics for the discovery.
[13] On the significance of the fact for both cosmology and cosmogony see "Relict radiation, infinity and the horizon" by A. D. Chernin, *Priroda*, No. 3, 1979.

ponding Member of the USSR Academy of Sciences; they worked with the RATAN-600 radio telescope.

Relict photons have been discovered; now what about relict neutrinos, their close relatives and partners? It is practically impossible to register them directly, even while employing the most advanced modern experimental equipment: they interact too weakly with any other particles and therefore with any physical instrument. They do not possess any electrical charge and therefore do not take any part in electromagnetic interactions. They are also insensitive to the strong interaction. Only weak interactions and gravitation exist for them. However, neutrinos cannot be registered using the effects related to weak interactions. Now what about gravitation? Here there are grounds for hope. If the neutrino has a rest mass (this is an open question yet), they might be the enigmatic dark mass mentioned above. The concentration of these particles in the Universe is such that if their rest mass were about 20–30 electron volts (approximately thirty thousand times less than the electron mass), the related average density would correspond to the hidden mass. If this is so, the principal mass of the Universe belongs to the relict neutrinos. It is clear now that the problem is both exciting and fundamental ...

The discovery of relict radiation changed many aspects of cosmology. Alternative theories, for instance, the steady-state theory, were gradually put aside. Skeptical voices died down: even the most uncompromising skeptics no longer risk being scornful about cosmology. One of the erstwhile critics of relativistic cosmology remarked: "Whoever would have thought that such important and, more to the point, observable astronomical results could follow from such an abstract and meaningless theory." Here is another characteristic remark: "Cosmology has grown to become a respectable science."

Fifteen years passed from the prediction of relict radiation to its discovery. Gamow was luckier than his teacher: he lived to see the triumph of cosmology and the astounding success of the science to which he undoubtedly made the greatest contribution after Friedmann.

The hot-Universe theory opened up new horizons in cosmology and indicated new approaches to the most fundamental problem of cosmology and of science generally: the origin of the Universe. The key needed to unravel the enigma of the world's origin is kept by the relict particles: by photons, neutrinos, antineutrinos and other elementary particles (in particular, those with a mass much greater than that of the proton) which existed in the Universe during the first moments of its existence, some remaining until the present epoch, some disappearing but leaving their traces in basic features of the contemporary world, such as the rate of its expansion, its average density and the type of space geometry.

The elementary particles obey the laws of quantum mechanics. On the other hand, the Universe itself at the beginning of its expansion was hardly greater than a point, so that it was, as it were, an elementary particle then. But if so, the Universe as a whole must then have been governed by quantum laws. These laws controlled processes while the matter density was by many orders of magnitude greater than the density of an elementary particle such as the proton. In addition, the gravitational fields were immensely strong under those conditions, or, which is the same, space–time was curved enormously. Further investigation of those completely abnormal properties of the early Universe needs a theory covering both the physics of the smallest natural bodies, i.e. elementary particles, and the physics of the largest object, the Universe itself. Such a synthesis of microcosmic theory and cosmology has not yet been reached. However, the advance of physics during recent years has been strongly directed towards this aim, and reasons motivating the development of a unified theory arise not only in cosmology, but also from the intrinsic needs of the physics of elementary particles. Many physicists suggest there is a possibility that the physical processes occurring in the early Universe not only predetermined its own properties but also the properties of the elementary particles as they exist in this world, e.g., their masses and the pattern of their interaction.

This field of research has become the front line of fundamental physics. Among the profound and original ideas offered in the subject, we would like to draw attention to one advanced by Moisei A. Markov, Full Member of USSR Academy of Sciences, on some hypothetical elementary particles that can be regarded as entire universes almost completely locked in themselves. If they were closed completely, they would never exist for us at all. However, the closed state can be incomplete (e.g., if the general electric charge of the world is not zero but equals, say, the charge of an electron). Then such universes will take part in the activities of our physical world looking to us like elementary particles. Here is where we can recollect a couple of lines from Valery Bryusov, a Russian poet.

> Who knows, who, maybe these electrons
> Are worlds with continents and oceans.

These hypothetical particles are called friedmons in honor of the founder of relativistic cosmology.

One more idea has to be mentioned here because it is appealing in its scientific daring. Professor Pyotr I. Fomin suggested research on the following alternative: why couldn't the birth of the Universe be a quantum process of spontaneous creation from vacuum? Indeed, the birth of individual elementary particles from vacuum is a well-known fact in physics, it has been investigated in detail and proven by experiment. And if the Universe was once, as has been mentioned, like an elementary particle, why could it not be born in the way that is possible and even natural for an elementary particle? This idea has been put forward independently by other researchers as well, and it attracts the attention and efforts of the most resourceful and authoritative theorists.

If what we know or suppose concerning the early Universe is true, its history comprises the following stages or events. First, its quantum birth during an unimaginably small period of $c.$ 10^{-43} s, the initial size being $c.$ 10^{-33} cm (compare with the size of the proton: $c.$ 10^{-13} cm). Then it expanded for $c.$ 10^{-35} s with a gigantic gain in size to $c.$ 10^{106} cm (compare with the size of the observed region of the Universe: $c.$ 10^{28} cm). Until this moment, the numbers can be regarded as given just for the sake of illustration, as Friedmann might have said. However, the next stage, that of the hot state of the helium synthesis for 1–10 minutes has been investigated with greater reliability; practically no hypotheses are needed and exact theoretical analysis and computations are called for. The next stage is expansion accompanied by both cooling of matter and radiation. One to three billion years after the start of expansion, there begins the formation of major structures, i.e. galaxies and their systems and the first stars. In 10–15 billion years, among the new generation of stars in our Galaxy appears the Sun, and afterwards the planets around it. The birth of the Earth 4.5 billion years ago started evolution of another kind, bringing about the emergence of life and then of the human mind: possibly there is no other similar mind in the whole Universe. The human mind now strives, as Friedmann said, "to number the stars or in some other way ... to re-create the picture of the world on the basis of always infinitesimal scientific data."

The greatest progress in the field was achieved by Ptolemy, Copernicus, Galileo, Newton, Einstein, and Friedmann.

Conclusion

The Universe we live in can be described by a theory created by Alexander A. Friedmann. This theory is a triumph of scientific thought. For the first time ever the human mind embraced the Universe as a whole in its dynamics and development.

What was it that brought him such extraordinary success?

The young school mathematician was noticed by Markov; Steklov became his teacher at university; and in his last days one of his concerns was to publish a posthumous memoir of Lyapunov's. Friedmann was educated by the famous St. Petersburg school of mathematicians that had been founded by Chebyshev and made famous by those four outstanding scientists. His research displayed the excellent inherent features of the school: the ability to link mathematical problems with the fundamental issues of natural science, concrete choice of the object of research, sufficient generality in the problem set-up and a tendency to reduce the problem "to a number," making it possible to apply practically and verify experimentally the elaborated theory. Just as in mechanics and meteorology, so in cosmology he "worked it out to a number" and gave an estimate for the age of the Universe: 10 billion years as the order of magnitude.

He also absorbed the ideas of modern physics while studying it under the guidance of great scientists, e.g. Ehrenfest, one of the major theoretical physicists of the 20th century. Friedmann was able to receive from him both concrete knowledge and an independent, critical approach to any scientific problem, however well it might be supported by the authority of the classics.

Friedmann's broad general outlook, free of any dogmatism, born from the spirit of the dramatic events occurring both in his country and in the world, was nourished by the fruits of world culture from Indian myth to

254

the Bible, from Aristotle, Augustine and Dante to the advanced ideas and achievements of his own time.

He arrived in the 1920s as a mature scientist at the peak of his talent, a person whose lifestyle predisposed him to formidable effort. His scientific daring was supported by his personal valor and courage shown on numerous occasions in his eventful and far from secluded life: when he was young, he was among the leaders of the revolutionary year of 1905, during World War I he was an air reconnaissance officer, and in the last year of his life he became a fearless and record-breaking aeronaut. He thought of himself as a meteorologist and was a recognized authority in the subject, but he started to deal with the general theory of relativity and cosmology, both of which were greatly intriguing everybody. And he was fully prepared to take the seemingly easy step from large-scale motions in the atmosphere to dynamics on a cosmic scale. He introduced both motion and development firmly into the science of the Universe, became the founder of evolutionary cosmology, and overcame and destroyed the centuries-old paradigm of the static nature of the Universe that held back the thought of such innovating scientists as Einstein, Hilbert and Weyl.

Cosmology after Friedmann continued to develop his ideas, and the most important achievement in the field was the concept of the early "hot" Universe advanced by Gamow, Friedmann's student at St. Petersburg University. This theory was brilliantly confirmed by the 1965 discovery of the relict radiation it had predicted.

"So far Friedmann's name has been unduly forgotten. This is unjust, and must be corrected. We must leave his name in history," said P. L. Kapitsa, a Full Member of the USSR Academy of Sciences, when he spoke at the Academy on the celebration of the 75th anniversary of Friedmann's birth.

During the years that have passed since then, much has been done to study and develop Friedmann's legacy. Academician Ya. B. Zeldovich has put much energy into it; his books and articles, exemplary in their clarity and depth, have made Friedmann's work familiar to many scientists both in the USSR and in the world at large. Both Friedmann and Zeldovich got involved in cosmology after becoming authorities on hydrodynamics. Zeldovich has also made a decisive contribution to the formation of a new initiative in research, tying up cosmology and the physics of elementary particles. The most important goal of this initiative is to study the problem of the origin of the Universe, which was posed by Friedmann.

Friedmann's ideas were to have a happy destiny. His fame is expanding. And his cosmology is increasingly perceived as a fundamental element of science and human culture.

Main dates in Friedmann's life and work

Alexander Alexandrovich Friedmann was born on June 4 (16), 1888 in St. Petersburg.

1897–1906   Student at the 2nd St. Petersburg (Alexander I) Gymnasium
1905–   First Scientific study written (published in 1906)
1905–06   Takes part in the students' movement of St. Petersburg secondary schools; member of the Central Committee of this organization
1906–10   Student at the Faculty of Physics and Mathematics of St. Petersburg University
1910–13   Retained at the University to prepare for professor's work
1910–14   Teaching practical classes in mathematics; lecturer at the Institute of Railway Engineering
1912–14   Lecturer at the Institute of Mining
1913   Passes Master's examinations and becomes a Master's degree student at St. Petersburg University
1913–14   Staff worker at the Nicholas Aerological Observatory in Pavlovsk (a suburb of St. Petersburg)
1914   Sent to Leipzig (Germany), scientific research work under V. Bjerknes
1914–16   Takes part in World War I; volunteers to join the Army, serving in aviation units of the northern and southern fronts
1916–17   In charge of the Central Aeronautical Service (CAS) of the front; instructor at the Kiev school of observer-pilots
1917–18   Employee, and later Acting Director, of the Moscow "Aviapribor" plant under the Air Force Administration
1918–20   Professor of the Chair of Mechanics at Perm State University
1920–24   Research worker in the Atomic Commission at the State Optical Institute
1920–24   Teaching mathematics and mechanics at the Faculty of Physics and Mechanics of Petrograd University
1920–25   Professor of the Faculty of Physics and Mathematics of the Petrograd Polytechnical Institute
1920–25   Instructor, and from 1921 Professor, in the Department of Applied Aerodynamics in the Faculty of Air Communications at the Petrograd Institute of Railway Engineering
1920–25   Junior scientific assistant, senior supervisor of studies at the Naval Academy
1920–25   Senior physicist, head of the mathematical bureau, learned

secretary and from February 1925 Director of the Main Physical (later Geophysical) Observatory

1922 Writes and publishes the article "On the curvature of space"

1922 Defends his Master's dissertation; the corresponding book *The Hydromechanics of a Compressible Fluid* is published

1923 The book *The World as Space and Time* is published

1923–25 Editor of the journal *Geophysics and Meteorology*

1923 Scientific mission to Germany and Norway

1924 Attends the International Congress for Applied Mechanics in Delft (Netherlands), elected member of the standing committee for convening the Congress

1924 First volume of the book *Fundamentals of the Theory of Relativity* is published (written jointly with V. K. Frederiks)

1925 Editor-in-chief of the Journal *Klimat i Pogoda* (Climate and Weather)

1925 Edits the geophysics section of the great Soviet Encyclopedia

July 1925 Ascends to a record-breaking altitude of 7,400 m in a balloon and conducts investigations there (jointly with P. F. Fedoseyenko)

On September 16, 1925, Alexander Alexandrovich Friedmann died in Leningrad. Buried (on September 18) in the Smolensky cemetery.

Bibliography

General

Friedmann, A. A. *Selected Works* (in the series *Classics of Science*), Nauka
Publishers, Moscow, 1966, 462pp. (Especially: Friedmann's autobiography;
his letters to B. B. Golitsyn and V. A. Steklov; articles by V. A. Steklov,
Ye. P. Friedmann, A. F. Gavrilov, and L. S. Polak; and, of course,
Friedmann's papers.)

Collections of papers commemorating Friedmann

Geofizichesky Sbornik (Geophysical Collection), issue 1, 1927; issue 2, 1928;
issue 3, 1932
Zhurnal Geofiziki i Meteorologii (Journal of Geophysics and Meteorology),
vol. 2, 1925; vol. 3, 1926
Klimat i Pogoda (Climate and Weather), Nos. 1–3, 1925
Trostnikov, V. N. *Outstanding Soviet Scientist A. A. Friedmann*, Znanie
Publishers, Moscow, 1963, 48pp.

Archives

State Historical Archive
State Historical Archive of the October Revolution and Socialist Construction
State Historical Archive of Leningrad
Archive of the USSR Academy of Sciences (Leningrad Department)
Archive of the M. I. Kalinin Leningrad Polytechnical Institute

Chapter 1

Grigoriev, S. *A Historical Reference. Cantonist Berko*, Moscow & Leningrad,
1941, pp. 161–4

Chapter 2

Dianin, S. A. *Revolutionary Youth in Petersburg in 1897–1917*, Priboi Publishers, Leningrad, 1926, 244pp.

Chapter 3

Ignatsius, G. I., *Vladimir Ignatievich Steklov*, Nauka Publishers, Moscow, 1967, 211pp.

Vladimirov, V. S., and Markush, I. I. *Vladimir Ignatievich Steklov as a Scientist and Organizer of Science*, Nauka Publishers, Moscow, 1981, 95pp.

Andreyev, N. N., "My meetings with Friedmann," *Priroda* (Nature), 1967, No. 6, 88–93

Chapter 4

Krylov, A. N., *Collected Works*, vol. 1, part 1, AN SSSR Publishers, Moscow & Leningrad, 1951, 304pp.

Budyko, M. I., "The Main Geophysical Observatory and the advance of domestic meteorology," in *Zemlya i Vselennaya* (Earth and Universe), No. 5, 1967, 60–5

Khramov, S. P. "Weather Tomorrow," in *Zemlya i Vselennaya* (Earth and Universe), No. 3, 1967, 14–22; No. 4, 11–16

Chapter 5

Strokov, A. A. *Armed Forces and Military Art in the First World War*, Voenizdat Publishers, Moscow, 1974, 616pp.

Bogolyubov, A. N., "The Kiev Physico-Mathematical Society." In *The History of Mathematical Science*, Naukova Dumka Publishers, Kiev, 1984, pp. 3–7

Chapter 6

A. M. Gorky Perm State University. A Historical Essay. 1916–1966, Perm, 1966, 291pp.

Loskutov, K. N., "A. A. Friedmann's Activities at Perm University," in *Scientific Notes of A. M. Gorky Perm State University*, No. 163, Physics, Perm, 1968, pp. 149–56

Chapter 7

Kibel, I. A., "Hydrodynamical (numerical) short-term weather prediction," in *Mechanics in the USSR for 50 Years*, vol. 2, Nauka Publishers, Moscow, 1970, pp. 561–583

Kochin, N. Ye., *Collected Works*, vol. 1, AN SSSR Publishers, Moscow, 1970, pp. 561–83

Monin, A. S., and Yaglom, A. M., *Statistical Hydromechanics (the Mechanics of Turbulence)*, part 1, Nauka Publishers, Moscow, 1965, 640pp.
Serrin, J., *Mathematical Principles of Classical Fluid Mechanics*, Handbuch der Physik vol. 8/1, Springer-Verlag, Berlin, 1959

Chapter 8

Vizgin, V. P., and Frenkel, V. Ya. "Professor Vsevolod Konstantinovich Frederiks (on the centenary of his birth)." In *Einstein Collection 1984–1985*, Nauka Publishers, Moscow, 1988
Weyl, H., *Raum – Zeit – Materie*, Berlin, 1921
Hilbert, D., "Die Grundlagen der Physik," *Nachrichten der Gesellschaft der Wissenschaften Göttingen*, 1915, 395–407 (part 1), and 1917, 53–76 (part 2)
Einstein, A., *Collection of Selected Works*, vols. 1 & 4, Nauka Publishers, Moscow, 1965, 1967
V. I. Lenin's Library in the Kremlin. Catalog, Politizdat Publishers, Moscow, 1961, 763pp.

Chapter 9

Albert Einstein and the Gravitation Theory. Collection of Articles, Mir Publishers, Mowcow, 1979
Misner, C. W., Thorne, K. S. and Wheeler, J. A. *Gravitation*, vols. 1–3, Mir Publishers, Moscow, 1976 (first edn. W. H. Freeman, San Francisco, 1973)
Frenkel, V. Ya., "New materials on the discussion between Einstein and Friedmann on relativistic cosmology," in *Einstein Collection 1973*, Nauka Publishers, Moscow, 1974, pp. 5–18
Frenkel, V. Ya., "Einstein and Soviet physicists," *Voprosy Istorii* (Problems of History), No. 3, 1976, 25–30
Tatarinov, Yu. B., "Sixty years of the expanding Universe theory," in *Voprosy Istorii* (Problems of History), No. 3, 1982, 88–97

Chapter 10

Gulo, D. D., and Osinovsky, A. N., *Dmitry Sergeyevich Rozhdestvensky*, Nauka Publishers, Moscow, 1980, 281pp.
Loitsyansky, L. G., and Lurie, A. I., "Alexander Alexandrovich Friedmann (1888–1925)," *Proceedings of the M. I. Kalinin Leningrad Polytechnical Institute*, No. 1, 1949, 83–6
Devyatkova, Ye. D., *Recollections about A. F. Ioffe*, Nauka Publishers, Leningrad, 1973, pp. 70–84

Chapter 11

Fedoseyenko, P. F., "At an altitude of 7,400 m," *Vestnik Vozdushnogo Flota* (Bulletin of the Air Fleet), No. 9, 1926
Friedmann, A. A. "At an altitude of 7,400 m," *Klimat i Pogoda* (Climate and Weather), No. 2–3, 1926, 8–11

Yudin, M. I., and Kogan-Beletsky, G. I., "Alexander Alexandrovich Friedmann (on the 75th anniversary of his birth)," *Meteorology and Hydrology*, No. 12, 1963, pp. 43–5

Chapter 12

Shafirkin, V., "On the structure of the Universe and some reactionary ideas of bourgeois cosmology," in *Pod Znamenem Marksizma* (Under the Banner of Marxism), No. 7, 1938, 115–36

Lvov, V. Ye., "At the front of cosmology," ibid., No. 8, 1938, 137–67

Eigenson, M. S., "A crisis of bourgeois cosmology," in *Priroda* (Nature), No. 7, 1950, 12–18

Seleshnikov, S. I. *The Struggle of Materialism against Idealism in Astronomy*, Leningrad, 1956, 43pp.

Zeldovich, Ya. B., "The theory of vacuum may solve the enigma of cosmology," in *UFN*, vol. 133, 1981, p. 479

Gorelik, G. Ye., and Frenkel, V. Ya., "Matvei Petrovich Bronshtein," in *Einstein Collection 1980–1981*, Nauka Publishers, Moscow, 1985, pp. 291–327

Linde, A. D., "The evolving Universe," in *UFN*, vol. 144, 1984, p. 177

Longair, M. S. and Einasto, J. (eds.), *The Large-scale Structure of the Universe*, Mir Publishers, Moscow, 1981 (first edn. Reidel, Dordrecht, 1978)

Grib, A. A., Mamayev, S. G., and Mostepanenko, V. M. *Quantum Effects in Intense External Fields*, Moscow, Atomizdat, 1981, 295pp.

Zeldovich, Ya. B., and Novikov, I. D., *The Structure and Evolution of the Universe*, Nauka Publishers, Moscow, 1975, 735pp.

Name index

Abel, N. H. 42
Ackeret, J. 190
Adamov, A. A. 49, 181–2
Afanasiev, A. P. 12
Afanasieva, V. I. 98
Afanasieva-Ehrenfest, T. A. 34, 39n, 191
Alembert, J. de R. d' 139
Alexander I, Emperor 13, 62
Alexander, Grand Duke 69, 71, 81
Alexandrov, G. F. 223
Alin, N. I. 89
Alpher, R. 248–9
Ambartsumian, V. A. 245
Amundsen, R. 200
Andreyev, K. A. 21
Andreyev, N. N. (meteorologist) 18, 46, 77, 86–7, 259
Andreyev, N. N. (physicist) 46n
Anufrieva (Friedmann), L. Ya. 6
Appell, P.-E. 21, 32, 48, 78
Aristotle 132–3, 255
Arnold, V. V. 121
Arsenieva-Geil, A. N. 245
Auerbach, F. 120
Augustine, Saint 135, 255

Bachmann, 21
Backlund, O. A. 75–6
Bastilov, S. A. 198
Baumgardt, K. K. 12, 34, 177
Belodubrovsky, Ye. B. ix, 18
Belopolsky, A. A. 221
Beltrami, E. 110
Bergson, H. 121
Berker, R. 110
Bernoulli, D. 77
Bernoulli, J. 21, 33, 44
Bertrand, J.-L.-F. 21
Besicovich see Bezikovich
Bestuzhev-Riumin, K. N. 34, 55

Bezikovich, A. S. 39, 45, 47, 49–50, 89, 91–2, 94–5, 181
Bezikovich, Ya. S. 181
Białobrzeski, Cz. T. 84
Bianchi, L. 21, 48, 163
Bitov, M. 91
Bjerknes, J. 99
Bjerknes, V. F. 59, 63, 99, 189, 213, 256
Blinova, Ye. N. 103, 214
Blumenthal, L. O. von 192
Bobylev, D. K. 21, 29, 43–5, 56, 83, 111
Bogdanov, Ya. B. 40
Bogolyubov, A. N. 259
Bogoyavlensky, S. 4
Bohr, H. A. 50
Bohr, N. S. 190–1
Boltzmann, L. 34, 56n
Bolyai, J. 123
Bondi, H. 222, 243
Borel, E. 21, 38
Borisyak, A. 37
Born, M. 121
Brewer, E. C. 42
Brockhaus, F. A. 2, 6, 36
Bronshtein, M. P. 221–3, 225, 261
Bruno, G. 146
Bryusov, V. Ya. 252
Bubnov, I. G. 59–60
Buchstab 211
Budyko, M. I. 62, 259
Bukreyev 21
Bulgakov, N. A. 29
Bulygin, V. V. 43–5, 47, 49, 51, 56–7, 69, 83, 181
Bunyakovsky, V. Ya. 21
Bursian, V. R. 34, 116, 182, 190, 221, 245

Cantor, G. 21
Cassirer, E. 120–1
Cauchy, A. L. 21, 42
Chaliapin, F. I. 177

Chapelon, J. J. 210
Chaplygin, S. A. 54
Chebyshev, P. L. 21, 28, 254
Chernin, A. D. ix, 137, 224n, 233n, 247n
Chernyshov, A. A. 182
Christyakov, V. M. 200
Chugaev, L. A. 30
Chukovsky, K. I. 186, 209
Chulanovsky, V. M. 34, 190
Clausius, R. J. E. 21
Clebsch, R. F. A. 48
Copernicus, N. 146, 153, 175, 253
Coriolis, G. G. de 66, 107
Coulomb, C. A. de 59
Courant, R. 192
Crelle, A. L. 33
Crocé-Spinelli, J. E. 206

Dante 175, 193, 255
Davidenkov, A. I. 13–14, 16–17, 20, 22, 25
Davidenkov, N. N. 13
Davidenkov, S. N. 13
Dedekind, R. J. W. 21
Defant, A. 99
Delone, B. N. 84, 180
Delone, N. B. 84
Delyanov, I. D. 111
Demianski, M, x
Devyatkova, Ye. D. 184, 260
Dianin, S. A. 22, 26, 42, 43n, 45n, 259
Dicke, R. 246
Dirichlet, P. G. L. 21
Dobrolyubov, N. A. 28
Doinikova, V. V. 8, 11, 32, 34–5, 37–9, 89, 94, 187, 193
Dollond, J. 76
Doppler, C. J. 217
Dostoyevsky, F. M. 37
Dreyer, J. L. E. 217
Droste, J. 172
Du Bois-Reymond, P. D. G. 21
Dushevsky, Ye. P. 198, 202
Dymnikova, I. G. 241

Eddington, A. S. 159–61, 164, 217, 219, 221–2, 236–7, 241–2
Efron, I. A. 2, 6, 36
Ehrenfest, P. S. 1–2, 11, 34–43, 56n, 57, 95, 171–2, 177–8, 191–2, 254
Ehrenfest, T. A. *see* Afanasieva-Ehrenfest
Ehrenfest, T'. 36, 38
Eigenson, M. S. 261
Einasto, J. E. 233, 238–9, 261
Einstein, A. 48, 54, 103, 114, 120–2, 138–42, 144–53, 155, 157, 159, 160–4, 167, 168, 169–74, 190, 209, 215–16, 219, 221–3, 236–8, 240–1, 253, 255, 260
Emden, R. 100

Engberts, A. E. 35n
Engelke, V. 20
Eriksen, J. 110
Euler, L. 21, 57, 66
Everding 213

Feddersen, B. W. 47
Fedorov, I. 4
Fedoseyenko, P. F. 202–7, 212, 257, 260
Feringer, A. B. 34, 189
Fermi, E. 248
Fersman, A. Ye. 196
Ficker, H. 72–3, 188–9, 210
Flettner, A. 190
Flit, P. P. van der 54
Fock, V. A. 98, 115, 162–3, 172, 180, 190–1, 224–5, 245
Fomin, A. 41
Fomin, P. I. 253
Fourier, J.-B. J. 57
Frederiks, V. K. 10, 115–17, 120, 171, 178, 180, 182, 190, 213, 221, 223, 245, 257, 260
Frenkel. V. Ya. ix, 2, 8–11, 49, 54n, 191n, 222n, 260–1
Frenkel, Ya. I. 10, 115, 121, 245
Freundlich, E. F. 120, 174
Friedmann, A. A. (father) 2, 3, 4–5, 7–8, 12, 24, 29, 41, 193
Friedmann, A. A. (father's stepbrother) 6
Friedmann, A. A. (son) 9, 209
Friedmann, A. I. (grandfather) 4–6, 13, 27, 41, 193
Friedmann, L. A. (uncle) 5
Friedmann, M. A. (aunt) 4, 5, 8, 193
Friedmann, Ye. N. (grandmother) 5
Friedmann, (Dorofeyeva) Ye. P. (1st wife) 22, 55, 73, 78, 175, 186, 193, 258
Frolov, A. P. 7
Frunze, M. V. 192
Fuchs, I. L. 55–6

Galerkin, B. G. 60
Galileo 28, 139, 215, 253
Gamow, G. A. 242–51, 255
Gardenin, M. S. 84
Gauss, C. F. 42, 62, 112, 123, 227
Gavrilov, A. F. 10, 31, 33, 40, 45, 47–8, 51, 55, 60–1, 69, 77, 79, 84, 87, 90–1, 114, 177, 183, 185, 187, 258
Gelfond, A. O. 245
Gernet, N. N. 180
Gersevanov, N. M. 59
Ginzburg, V. L. 225
Glazenap, S. P. 17
Gliner, E. B. 241
Glinka, I. V. 16–17, 22, 25
Gold, E. 100, 213

Gold, T. 243
Golitsyn, B. B. 1, 59, 61, 63, 68, 74–5, 97, 258
Goltsman, M. I. 211
Gorelik, G. Ye. 222n, 261
Gorky, A. M. 179
Goursat, E. 21, 38, 48
Grave, D. A. 73, 84
Green, D. 107
Grib, A. A. ix, 261
Grigoriev, S. 258
Grimm, E. D. 88
Grinberg, G. A. ix, 175, 183
Grommer, J. 245
Grotz, K. 239
Gulo, D. D. 260
Gunter, N. M. 29, 52–4, 61, 69, 89, 180–1, 191–2, 194, 213
Gurevich, L. E. 224n, 226, 241, 247n
Gurevich, Ya. Ya. 41
Guth, A. H. 242

Halley, E. 42–3
Heine, H. E. 48
Heitler, W. H. 245
Heller, M. 240
Helmholtz, H. L. F. von 104–6, 110–11
Hensel, P. N. 16, 18
Hergessel, G. 188–9, 191
Herman, R. 248–9
Hertz, H. R. 47
Hertzen, A. I. 62
Hesselberg, I. 189
Hesselberg, P. 189
Hesselberg, T. 63–7, 189, 213
Hessen, Ya. M. 196
Hilbert, D. 21, 34, 123, 190, 255, 260
Hoffmann, E. T. A. 37
Holtsmark, J. P. 63
Hopf, E. 112
Horn, J. 164
Hoyle, F. 243
Hubble, E. P. 216–21, 223, 229, 231–2, 236–7, 239, 250
Humason, M. 221
Humphreys, W. 100

Ignatsius, G. I. 259
Ilf, I. 15
Indrikson, F. N. 12
Iodynsky, Ya. V. 16, 18–19, 22
Ioffe, A. F. 34, 38, 57, 98, 181–2, 184, 190–2, 241, 246, 260
Ioganson, A. Kh. 8
Isakov, L. D. 34, 43
Itskovich, D. I. 212
Ivanenko, D. D. 245
Ivanov (Friedmann's classmate) 20

Ivanov, A. A. 56, 176
Ivanov, I. I. 29, 32, 40, 56, 182
Ivanov, V. V. ix
Izvekov, B. I. 98, 100, 102, 110, 192n, 199, 211–14

Jacobi, C. G. J. 33
Johannsen 7
Jordan, C. 21, 38
Jordan, P. 245
Josephson, P. 35n
Joukowski *see* Zhukovsky
Jover, M. 233n

Kagan, V. F. 42
Kalitin, N. N. 81, 201
Kalyandyk, S. I. 84
Kamenshchikov, N. P. 196
Kapitsa, P. L. 182, 255
Kapustin 20
Karayev, G. 18
Karazin, V. N. 62
Kardashev, N. S. 225
Karman, T. von 112, 192
Karpinsky, A. P. 211
Kassil, L. A. 15
Keller, L. V. 98, 111–13, 191, 212
Kelvin, Lord (Thompson, W.) 47, 87
Keppen, V. P. 188
Khaikin, S. E. 250
Khalatnikov, I. M. 225
Khramov, S. P. 259
Khvolson, O. D. 12, 29, 39, 94, 121, 167n, 221
Kibel, I. A. 65–6, 98–9, 103, 106–7, 110, 113, 214, 259
Kikoin, I. K. 175
Kirov, S. M. 2
Kirpichev, M. V. 182
Kirpichev, V. L. 9
Klapdor, H. V. 239
Klein, C. F. 34
Kneser, A. 210
Kobushkin, P. K. 225
Kochin N. Ye. 66, 98–9, 107–8, 110, 113, 191, 212, 214, 259
Kochina P. Ya. *see* Polubarinova-Kochina
Kogan-Beletsky, G. I. 261
Kolchak, A. V. 92–3, 96
Kolmogorov, A. N. 113
Kolyshkin, B. 22–3
Komarov, V. L. 31–2
Korkin, A. N. 28, 32, 57
Korovchenko, A. S. ix
Korrilov 204
Koshlyakov, N. S. 91
Kosonogov, I. I. 84
Kostareva, O. A. 98–9

Kotelnikov, A. P. 84
Koyalovich, B. M. 22
Krutkov, Yu. A. 34–5, 116, 170–3, 178, 180, 182, 190–1, 245
Krutkova, T. A. 171–3
Krylov, A. N. 30, 59, 62, 80–2, 182–3, 185, 190–1, 195, 259
Krylov, N. M. 54
Kupfer, A. Ya. 62
Kurchatov, I. V. 81
Kurnakov, N. S. 182
Kuzmin, R. O. 10, 89, 91–2, 181
Kuznetsov, N. 16

L. T. (Friedmann's pseudonym) 43n
Ladyzhenskaya, O. A. 49
Lagrange, J. L. 56
Lamé, G. 21, 58
Landau, L. D. 222, 224, 245
Laplace, P. S. de 44, 58
Laue, M. F. T. von 121, 164
Lazarev, P. P. 197
Lebedinsky, V. K. 47
Lee, T. 210
Legendre, A. M. 21, 42
Lehmann, I. 120
Lemaître, G. H. 219, 221–5, 236–7
Lempfert 108
Lenin, V. I. 121, 195, 260
Levi-Civita, T. 69, 131, 192, 210
Levitskaya, M. A. 34
Lichtenstein, L. 192
Lifshitz, S. 120–1
Lifshitz, Ye. M. 224–5, 246
Lilovy (Friedmann's conspiratorial name) 27
Linde, A. D. 242, 261
Linnik, V. P. 84
Liouville, J. 48
Lobachevsky, N. I. 123, 162
Loitsyansky, L. G. ix, 105, 112–13, 175, 183–4, 191, 212, 260
Lomonosov, M. V. 28, 63–4
London, Ye. S. 121
Longair, M. S. x, 261
Lorentz, H. A. 99, 171, 178
Lorenz, E. 64
Loris-Melikov, M. A. 212
Loskutov, K. N. 93, 259
Love, A. E. H. 58
Lowell, P. 217
Lucas de Pesloüan, Ch. 42
Lurie, A. I. 105, 183, 260
Lutsenko 77
Lvov, V. Ye. 261
Lyapunov, A. M. 28, 30, 33, 47, 210, 254

McCrea, W. H. 226–30, 234

Mach, E. 66
Machinsky, M. V. 225
Mai, K. I. 12
Malinina, N. Ye. (2nd wife) 173, 185–9, 208–9
Malinina, S. Ye. ix, 209
Mamayev, S. G. 261
Marchuk, G. I. 214
Maria, Empress 49
Markov, A. A. 21, 22, 28, 30, 38, 49, 56–7, 94, 254
Markov, M. A. 225, 252
Markush, I. I. 22, 259
Marquis 107
Mashkov, N. V. 91
Mashkovsky, A. 121
Mayakovsky, V. V. 186
Meinardus, W. S. 210
Melander, G. 210
Melgunov, S. P. 41
Meller, I. K. 7
Meller (Voyachek), O. I. 7
Merezhkovsky, D. S. 37
Meshchersky, I. V. 83, 103–4, 106, 182, 212–13
Milne, E. A. 224, 226–30, 234
Minkowski, H. 136
Mirolyubov, N. N. 193
Mises, R. E. von 174
Misner, C. W. 260
Mitkevich, V. F. 34
Molchanov, P. I. 81
Monin, A. S. 260
Morozov, N. A. 121
Mostepanenko, V. M. 261
Muskhelishvili, N. I. 190

Napoleon 11
Napravnik, E. F. 6
Nemchinov, V. S. 77
Newton, I. 59, 139–40, 150, 226–30, 234, 238, 244, 253
Nicholas II, Emperor 37
Nikolai, Ye. L. 182
Nordmann, C. 121
Novikov, I. D. 224n, 261

Oblakov, A. A. 4
Obolensky, V. N. 196
Obreimov, I. V. 92n
Obukhov, A. M. 113, 214
Odintsov, M. M. 203
Oldenburg, S. F. 196, 211
Orbeli, I. A. 13
Osinovsky, A. N. 260
Osinsky, V. 212
Ostriker, J. P. 238–9, 246
Ottokor, N. P. 93

Overbeck 107

Pahlen, E. A. von der 84, 174
Pariysky, Yu. N. 250–1
Pascal. B. 145
Pauli, W. 121
Pavlov, F. I. 42
Pavlovsky, N. N. 182
Peebles, P. J. E. 222, 238–9, 246
Penzias, A. A. 250
Pepov, (Ya. D. Tamarkin's conspiratorial
    name) 27
Perlitz, H. G. 34
Pestalozzi, J. H. 185
Petelin, M. F. 16, 33, 45, 47–51, 55, 59, 69,
    73, 80–1
Petri, R. J. 204
Petrov, F. N. 198
Petrov, Ye. 15
Picard, C. E. 21
Pius XII, Pope, 225
Planck, M. K. E. L. 87, 210
Poe, E. A. 37
Pogorzhalsky, A. 15
Poincaré, J. H. 21, 33, 47, 101
Poisson, S. D. 56
Pokrovsky, K. D. 88, 91
Polak, L. S. 258
Polenov, B. P. 88–9
Polubarinova-Kochina, P. Ya. 98–9, 107,
    113, 124, 212, 214
Pontecorvo, B. M. 225
Pound, R. V. 141
Prandtl, L. 112, 190
Ptashitsky, I. L. 29
Ptolemy 253
Pushkin, A. S. 12, 13, 189
Pythagoras 130, 154, 165

Quarenghi, G. 13

Rakipova, L. R. 103
Raman, C. V. 210
Rayleigh, Lord (Strutt, J. W.) 107–8
Rebka, G. A. 141
Rees, M. 246
Reynolds, O. 66, 99, 112
Riemann, G. F. B. 21, 123, 138, 142
Riemann, H. 2, 6
Rimsky-Korsakov, N. A. 2, 7
Ritz, W. 60, 97
Rocher, V. K. 84
Rossby, C. G. A. 110
Rozenkevich, L. V. 245
Rozental, I. L. 242n
Rozhdestvensky, D. S. 34, 177, 190–1, 260
Rubinstein, A. G. 6
Rubinstein, Ye. S. 211

Rukhovets, L. V. ix
Runge, C. D. T. 72
Rykachev, M. A. 61–3
Rynin, N. A. 54, 203, 212

Sakharov, A. D. 225
Santerre, S. 41–2
Savich, S. Ye. 29
Schmidt, H. 121
Schmidt, O. Yu. 84
Schoenflies, A. M. 21
Schrödinger, E. 57n
Schwarzschild, K. 100–1
Sedov, L. I. 66
Seleshnikov, S. I. 261
Selezneva, Ye. S. ix
Selivanov, D. F. 29, 32, 40, 56
Semyonov, N. N. 92n
Semyonov-Tien-Shansky, P. P. 80
Serret, J. A. 21
Serrin, J. 110, 260
Shafirkin, V. 261
Shakura, I. I. ix
Shaw, N. 107–8, 213–14
Shein, G. A. 89, 91
Shekhter, A. B. ix, 180, 184, 221
Shklovsky, I. S. 225, 248
Shmaonov, T. A. 250
Shokalsky, Yu. M. 197
Shokhat, Ya. A. 45, 47–8, 51, 53–7, 69, 73
Shuleikin, V. V. 197
Shvartzman, V. F. 169
Silk, J. 246
Sintsov, D. M. 73
Sitter, W. de 144–5, 151–3, 157, 159, 162–4,
    217, 221, 237–8, 241–2
Sivel, H. T. 206
Skobeltsyn, D. V. 182
Skobeltsyn, V. V. 182
Slipher, V. M. 217
Smirnov, D. A. 196–7
Smirnov, V. I. 10, 16, 27, 37, 45, 47–9,
    51–2, 55–7, 59, 61, 69, 73, 79, 91, 180–1,
    193, 213
Sokhotsky, Yu. V. 21, 29, 40, 57
Sokolov, F. F. 43
Sologub (Teternikov), F. K. 37
Sommerfeld, A. 190–1
Stachel, J. 173
Starobinsky, A. A. 241
Staropolsky 20
Steklov, A. I. 28
Steklov, V. A. 1, 28–34, 38, 40, 43–6,
    49–50, 52–7, 58–60, 68–73, 76–9, 87–91,
    94, 97, 101, 131, 176–7, 179–80, 183, 186,
    188, 191–3, 194–7, 201, 210, 254, 258–9
Steklova, O. N. 30, 68, 70, 72, 193
Stolz, O. 21

Strokov, A. A. 259
Sturm, J. C. F. 48
Suslov, G. K. 84
Sverdrup, E. 63
Syrtsov, A. I. 93

Tait, P. G. 87
Tamarkin, Ya. D. 15, 16, 17–22, 26–7, 29, 31–7, 39, 40, 43–5, 47–9, 52–6, 58, 60–1, 69, 73, 78–9, 83, 87, 96, 162, 176–7, 179–81, 191, 193
Tamarkina-Weichardt, Ye. G. 55
Tamm, I. Ye. 245
Tanakadate, A. 210
Tatarinov, Yu. B. 172, 260
Taylor, G. I. 103, 112
Teisserenc de Bort, L. P. 100
Teller, E. 244, 246
Thomson, W. *see* Kelvin
Thorne, K. S. 260
Tikhomirov, Ye. I. 81, 108, 192n, 198, 209, 211–12
Tikhomirov, P. K. 23, 25
Timoreva, A. V. 245
Tissandier, G. 206
Tolman, R. C. 216, 222
Tolstoy, L. N. 119n
Tomashevsky, V. B. 211
Tomilin, N. 42
Trapeznikova, O. N. ix, 180
Troitsky, S. N. 212
Tropp, E. A. ix
Trostnikov, V. N. 258
Tsiolkovsky, K. E. 203
Tudorovsky, A. I. 190, 245

Uspensky, Ya. V. 38, 73, 83, 181

Vajonyi, A. 110
Vasiliev, A. V. 121
Vasiliev, K. I. 212
Vasnetsov, M. K. 83n, 84
Vavilov, S. I. 120
Vinogradov, I. M. 89, 91

Vize, V. Yu. 200
Vizgin, V. P. 260
Vladimirov, V. S. 259
Voeikov, A. I. 10–11, 48n, 65, 82
Volterra, V. 60
Voronets, P. V. 84
Voskresensky, A. K. 46
Voyachek, I. K. (grandfather) 2, 6–7, 8
Voyachek (Friedmann), L. I. (mother) 2, 3, 7, 8–10, 193
Voyachek, V. I. (uncle) 2, 7, 8–10

Wagener, K. 187
Wangenheim. A. F. 198, 212
Weber, H. 21
Wegener, A. L. 188
Weichardt, G. G. 34, 55, 89, 91, 93, 95
Weickmann, L. F. 213
Weierstrass, K. T. 164
Weimer, H. 42
Weinberg, S. A. 226, 243–4, 249
Weyl, H. 67, 142–3, 218, 255, 260
Wheeler, J. A. 219, 260
Wien, W. 71
Wild, H. 62–3, 210
Wilson, R. W. 250
Witte, S. Yu. 9

Yaglom, A. M. 260
Yarilov, A. A. 200
Yavelov, B. Ye. 54n
Yefremidze, T. I. 191n
Yudin, M. I. 65, 102n, 214, 261

Zakharov, L. D. 204
Zeldovich, Ya. B. 225–6, 246, 255, 261
Zelmanov, A. L. 224–5
Zhdanov, A. A. 223–4
Zhitomirsky, O. K. 94
Zhukovsky, N. Ye. 54, 71, 84, 86–7
Zolina, Ye. M. 98
Zolotarev, Ye. I. 21
Zwicky, F. 232